A TEXTBOOK OF
NOVEL DRUG DELIVERY SYSTEMS

A TEXTBOOK OF
NOVEL DRUG DELIVERY SYSTEMS

Authors

Dr. V. Sankar

Professor and Head, Dept. of Pharmacy practice,
P.S.G college of Pharmacy, Coimbatore (Tamilnadu).

Dr. S. Ramesh

Associate professor and Head Dept. of Pharmaceutics,
Sri Padmavathi School of Pharmacy, Tirupati, A.P. (India).

Mr. V. Shanmugam

Asst. Professor in the Department of pharmaceutics,
Sri Padmavathi School of Pharmacy, A.P. (India).

PharmaMed Press

An imprint of BSP Books Pvt. Ltd.

4-4-309/316, Giriraj Lane,
Sultan Bazar, Hyderabad - 500 095.

A Textbook of Novel Drug Delivery Systems

by V. Sankar, S. Ramesh and V. Shanmugam

Tenth Reprint 2024

Published by

PharmaMed Press

An imprint of BSP Books Pvt. Ltd.

4-4-309/316, Giriraj Lane, Sultan Bazar, Hyderabad - 500 095.

Phone: 040-23445688; Fax: 91+40-23445611

E-mail: info@pharmamedpress.com

www.pharmamedpress.com/pharmamedpress.net

Printed at

Aditya Offset Process (I) Pvt. Ltd.

Hyderabad.

Price: Rs. 225.00

ISBN: 978-93-81075-96-8 (Paperback)

PSG COLLEGE OF PHARMACY
(An ISO 9001 : 2008 Certified Institution)
Affiliated to The Tamil Nadu Dr. M.G.R. Medical University, Chennai
Approved by Pharmacy Council of India and All India Council for Technical Education, New Delhi
POST BOX No. 1674, PEELAMEDU, COIMBATORE 641 004, TAMIL NADU, INDIA
2001-2011
Decennial Year
TÜVRheinland CERT ISO 9001
Phone : 0422-4345841
0422-2570170 (6 Lines)
Fax : 0422-2594400
E-mail : principalpsgcp@gmail.com
Website : www.psgpharma.ac.in

FOREWORD

Technological innovations have increased the scope of Pharmacy, particularly the drug delivery in the specialization of pharmaceutics. Novel drug delivery system is new fast developing branch aiming to deliver the drug through different dosage forms and targeting the specific tissues. Having this in mind the authors have written ten different chapters related to novel drug delivery. This book will bring further awareness and interest among the students to do research in pharmaceutics. Nanotechnology and its application in drug delivery is one of the most recent advancement in NDDS, this topic was written in lucid manner which will help the students. Special emphasize is also given to targeted drug delivery system in gastrointestinal system colon drug delivery systems.

Inclusion of other relevant topics like liposomes, microspheres will give the readers its significance in novel drug delivery. In all the chapters the authors have discussed different polymers used, preparation methods, characterization and formulation aspects. This novel approach on every drug delivery system will increase the interest on the subject and made easier for the readers.

The book also covers basic concepts in mucoadhesive drug delivery systems, oral controlled drug delivery systems, prodrugs, resealed erythrocytes and transdermal drug delivery systems. This book has been written with clear description, comprehensive figures along with illustrative examples. It will definitely be a good addition to the library and the readers will feel this book has value addition.

Dr. M. Ramanathan
Principal
Dr.M.Ramanathan,M.Pharm,Ph.D
Principal
PSG College of Pharmacy
Peelamedu, Coimbatore-4.

PREFACE

The first edition of this book entitled "A Textbook of Novel Drug Delivery System" consists of 10 chapters covering the preparation and evaluation of different novel formulations such as Nanoparticles, controlled drug delivery systems, Trans dermal patches etc.., It will be useful for the B. Pharmacy students and also for the M. Pharmacy students. The authors are highly thankful to the following members who were helped in the successful completion of this book.

Anusha. M, Bhargavi. T, Chandan Kumar Reddy. S, Deepak Peruru, Durga Srinivasa Rao. M, Keerthi Pratyusha. S, Madhu Sudhan. M, Mounika. CH, Yamini. K and Sai Sushma Priya.

Our scicere thanks to the Dr. D. Ranganayakulu, principal Sri Padmavathi School of Pharmacy, Tirupathi. Dr. M. Ramanathan, Principal PSG College of Pharmacy Coimbatore for their constant support and encouragement.

We express our gratitude to the management of Sri Padmavathi School of Pharmacy and PSG College of Pharmacy.

Above all we thank the almighty for the successful completion of the book.

Authors

CONTENTS

CHAPTER 1

NANOPARTICLES

Introduction

In recent years a prompt attention has been given in development of Novel Drug Delivery System (NDDS). Nanoparticle has occupied the most integrated part in novel drug delivery system. They are specially designed to release the drug in vicinity of target tissue.

- Speiser and his co-workers who first developed the concept that nanoparticles are used as vehicle for drug delivery in late 1960s and early 1970s.
- The word 'nano' refers to size ranging from 10-1000 nm and 'particle' is defined as a small object that behaves as a whole unit in terms of its transport and properties.

Definition

Nanoparticles may be defined as, a submicron colloidal systems generally, made up of polymers (biodegradable or not) with particle size ranging from 1-1000 nm, in which, the active ingredient is dissolved, entrapped or encapsulated or adsorbed or attached. They do not have precise classification but are referred as nanospheres, rods, fibers, capsules and suspensions.

Advantages

- Particle size and surface characteristics of nanoparticles can be easily manipulated to achieve both passive and active drug targeting after parenteral administration.
- They control and sustain the release of the drug during the transportation and at the site of localization, altering organ distribution of the drugs and subsequent clearance of the drugs, so as

to achieve increase in therapeutic efficacy of drugs and reduction in side effects.

- Controlled release and particle degradation characteristics can be readily modulated by the choice of matrix constituents.
- Drug loading is relatively high and drugs can be incorporated into the systems without any chemical reaction. This is an important factor for preserving the drug activity.
- Site-specific targeting can be achieved by attaching targeting ligands to surface of particles or use of magnetic guidance.
- The system can be used for various routes of administration including oral, nasal, parenteral and intra-ocular.
- Nanoparticles with diameter less than 200 nm are not screened out of circulation by liver and spleen.
- Nanoscale powders of anti-asthma and analgesic drugs are quickly absorbed in the human body in comparison to the traditional drug delivery systems.

Advantages of Nanoparticles Over other Drug Delivery System

An important area of application of nanotechnology includes novel drug delivery techniques, which are quicker and less risky, compared to the costs of developing new drugs.

- A large percent of ordinary drugs taken orally are destroyed by the stomach and liver and then distributed throughout the body, despite the fact that optimal effects would arise from focusing the drug to the target organ. Targeted drug delivery by nanoparticles has the potential to overcome some of these problems, and render treatment more effective, ensuring cost and safety benefits.
- The therapeutic potential of currently available drugs is also hampered by micro kinetics such as local instability issues and difficulties in crossing certain biological barriers such as blood brain barrier and placenta. Nanoparticles can help to address these problems, and several applications in drug delivery using this technology are being developed.

Limitations

In spite of these advantages, nanoparticles do have following limitations.

- For example, their small size and large surface area can lead to particle aggregation, making physical handling of nanoparticles difficult in liquid and dry forms.
- In addition, small particles size and large surface area readily result in limited drug loading and burst release.
- These practical problems have to be over comed before nanoparticles can be used clinically or made commercially available.

Preparation Techniques of Nanoparticles

Based on the physicochemical characteristics of polymer and drug to be loaded appropriate method for preparation is selected. The drug can be either entrapped within the reservoir or the matrix or otherwise be adsorbed on the surface of the particulate system. The classifications of methodologies are:

1. Macromolecule cross linking:
    (a) Heat cross linking
    (b) Chemical cross linking
2. Polymerization based methods:
    (a) Polymerization of monomer
    (b) Emulsion (micellar) polymerization
    (c) Dispersion polymerization
    (d) Interfacial condensation polymerization
    (e) Interfacial complexation
3. Polymer precipitation methods:
    (a) Solvent extraction/evaporation
    (b) Solvent displacement (nano precipitation)
    (c) Salting out

1. Macromolecule Crosslinking

Nanoparticles are prepared from amphiphilic (affinity for aqueous and lipid solvents) macro molecules, proteins and polysaccharides.

- The technique involves aggregation of amphiphilic followed by stabilization by heat denaturation or chemical cross linking.
- These process occur in biphasic either o/w or w/o dispersed system.

Cross Linking in W/O Emulsion

- This method is widely used for nano encapsulation of drugs.
- This method involves emulsification of bovine serum albumin (BVA), human serum albumin (HSA) or protein aqueous solution in oil under high pressure homogenization.
- This w/o emulsion is poured into pre heated oil (100°C) and stirred for a specific time to denature and aggregate protein contents of aqueous phase.
- Water is evaporated, proteinaceous sub-nanoparticles are formed which are washed with organic solvents and collected by centrifugation.

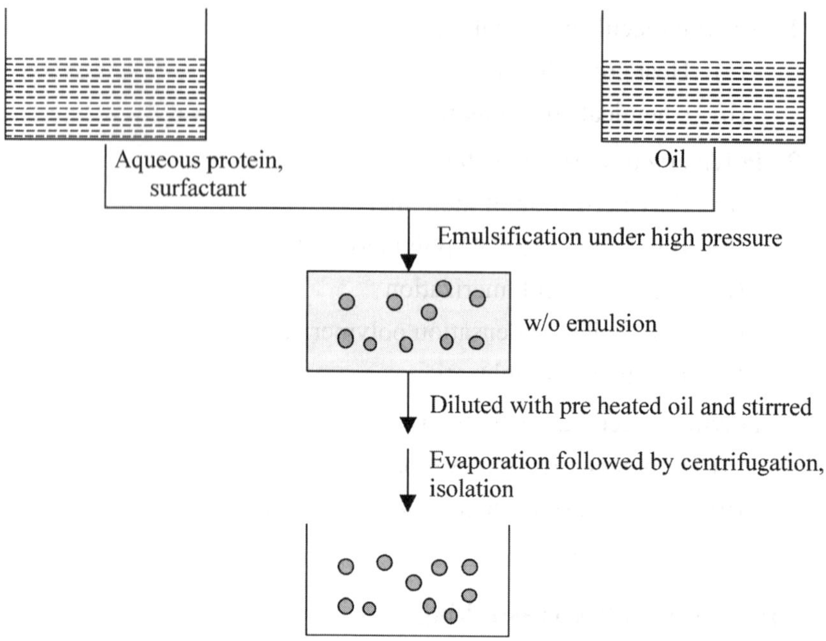

Fig. 1.1 Cross Linking in W/O Emulsion.

Phase Aeperation in Aqueous Medium (Desolvation)

- By changing PH or changing temperature or adding appropriate counter ions we can desolvate proteins or polysaccharides from aqueous phase.
- This method essentially proceeds by three steps
  - protein dissolution
  - protein aggregation
  - protein deaggregation
- All the three levels are maintained using appropriate levels of desolvation and resolvation and finally the aggregated particles are cross linked by using glutaraldehyde.

2. Polymerization based Methods

(a) Polymerization based Methods

- Two different approaches are adopted for preparation of nanospheres using polymerization technique.
- Emulsion polymerization: Non-solvent phase is used for the emulsification of the monomer to be polymerized.
- Monomer is dissolved in internal phase.
- Dispersion polymerization: Monomer is dissolved in a solvent in which polymer is immiscible.
- Monomer is taken as dispersed phase.

In both the cases polymer is insoluble in internal or dispersed phase resulting in a suspension of nanospheres.

(b) Emulsion Polymerization

- Depending on the nature of continuous phase the process of emulsion polymerization may be inverse or conventional.
- If continuous phase is aqueous (o/w emulsion) it is conventional. Whereas it is organic (w/o emulsion) it is inverse.

Two different mechanisms were proposed for emulsion polymerization.

(i) Micellar nucleation and polymerization

(ii) Homogenous nucleation and polymer

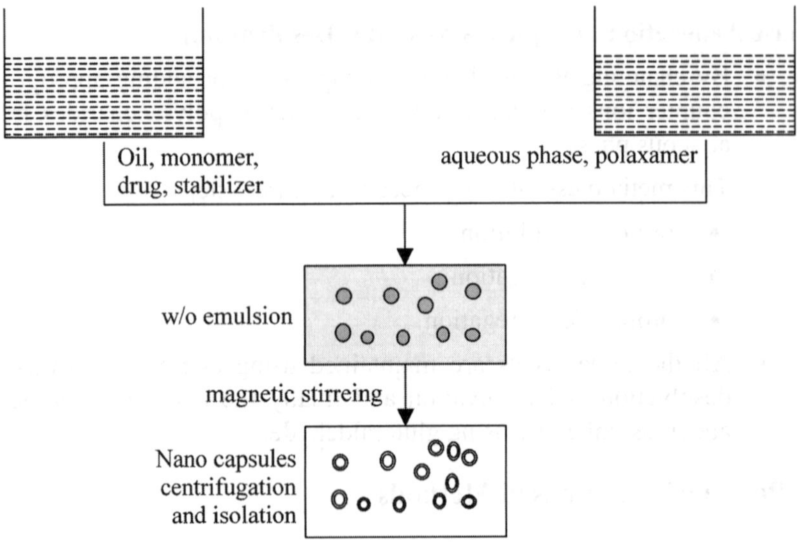

Fig. 1.2 Emulsion polymerization process.

(c) Dispersion Polymerization

- In this method the monomer is added to the aqueous phase which acts as precipitant for the polymer.
- So without catalyst only nucleation occurs in the aqueous monomer solution.
- The oligomers are formed and aggregated and above certain limit they precipitate as particles and transform into nanoparticles.
- This method is used to prepare biodegradable polyacrylamide and polymethyl-methacrylate nanoparticles.

(d) Interfacial Polymerization

- In this method polymer is preformed and finally transformed to an embryonic sheath.
- Polymer that forms core and drug to be loaded are dissolved in volatile solvents.
- Polymer and core phase are obtained by pouring the solution into non-solvent.

- At o/w interface the polymer phase is separated as a coacervation phase.
- Owing to the formation of nanocapsules the resultant mixture turns milky.
- E.g.: Interfacial polymeric condensation of 2, 2-bis-(4-hydroxy phenyl) propane and sebacoyl chloride.

(e) Interfacial Complexation

- In case of nanoparticle preparation with the help of surface active agents aqueous polyelectrolyte solution is carefully dissolved in reverse micelle in a polar bulk face.
- In soluble polyelectrolyte complex is formed by adding competing polyelectrolyte to coacervate at the interface.

3. Polymer Precipitation Methods

(a) Solvent Extraction

- Involves the formation of conventional o/w emulsion between partially water miscible solvent with polymer, drug and aqueous phase containing stabilizer.
- Solvent is removed by evaporation and addition of water.

Solvent Evaporation

- In this technique preformed polymer and drug are mixed with water immiscible organic solvent emulsified in a aqueous solution containing stabilizer.
- This crude emulsion is then exposed to ultra sonicator (a high energy source) or passed through homogenizers', colloid mills to reduce globule size.
- Fine dispersion of aqueous nanoparticles are formed by subsequent removal of organic solvent by heat and vacuum.
- Homogenization step is the key factor in obtaining submicron particles.

- E.g. PLGA (poly acetyl glycolic acid) to prepare PLGA nanospheres the polymer is solubilized in chloroform and dispersed in gelatin solution by sonication to form o/w emulsion.
- Solvent is removed by evaporation using sonicator, high-speed/pressure.
- Homogenizer which breaks the coarse emulsion to nano droplets yielding nanospheres.

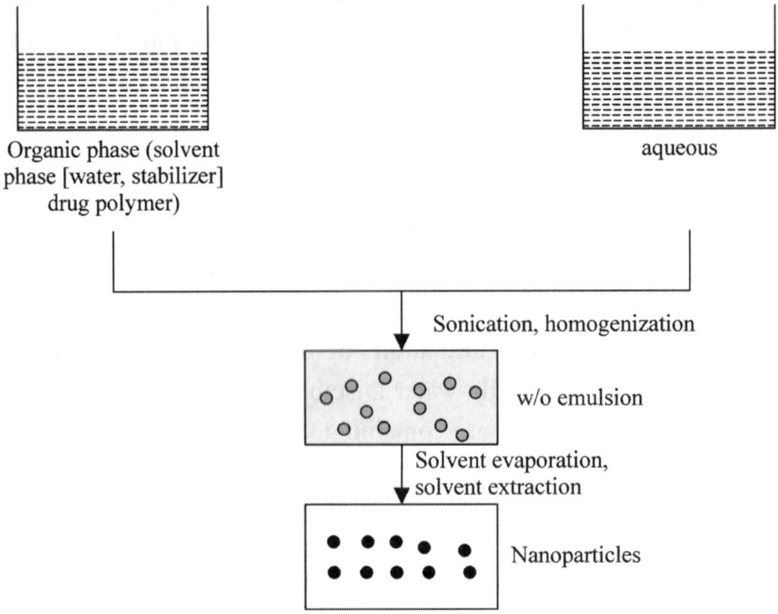

Fig. 1.3 Solvent evaporation method.

Multple Emulsion Solvent Evaporation Method

- It is modified form of solvent evaporation.
- In this method w/o/w is formed and later organic solvent is evaporated to form nanoparticles.
- This are washed with water and then lyophilized.

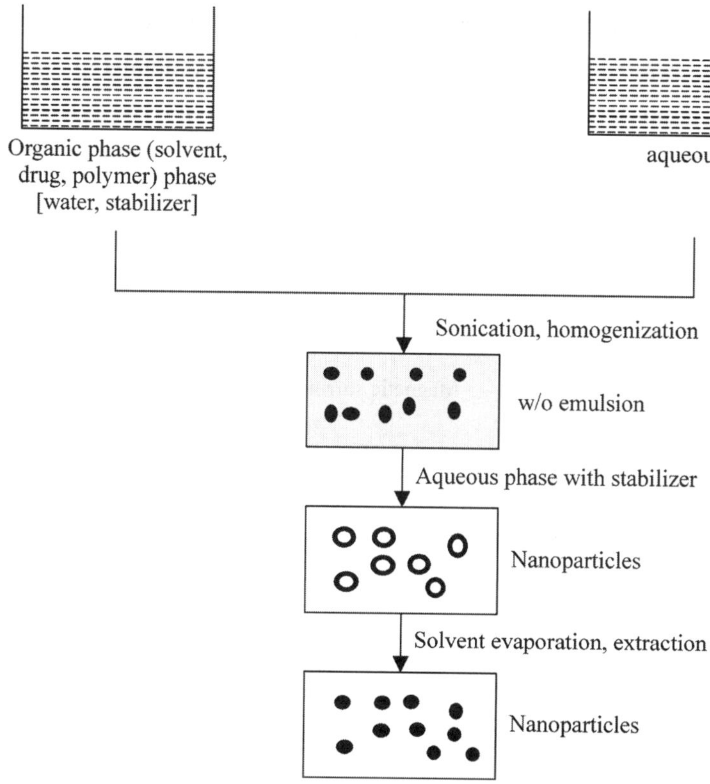

Fig. 1.4 Double emulsion solvent evaporation.

(b) Solvent Displacement (Nano Precipitation)

- This method involves interfacial deposition of polymer following displacement of semi polar solvent (miscible with water) from lipophilic solution.
- It involves use of organic phase and external phase which are completely miscible.
- As they are completely miscible the organic solvent diffuse to the external aqueous phase inducing polymer precipitation.
- For slightly soluble drugs after nano particle preparation solvent is eliminated and free flowing particles are obtained under reduced pressure.

- For hydrophilic drugs –drug diffuse out in to external phase.
- For highly hydrophobic drugs –drug precipitates in aqueous phase as crystals which grow during storage.

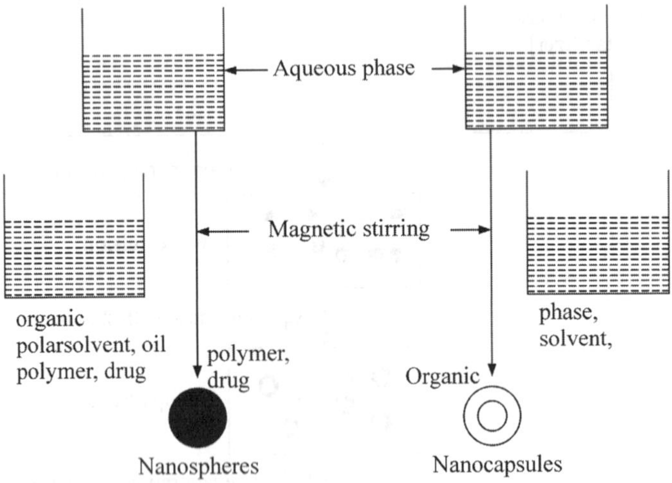

Fig. 1.5 Solvent displacement method.

(c) Salting Out

- Most common method used for preparation. It involves incorporation of saturated aqueous solution of poly vinyl alcohol (PVA) solution into organic phase.
- They are stirred under magnetic stirrer to form o/w emulsion.
- In this method miscibility of phases is retarded by adding PVA in external aqueous phase.
- When sufficient amount of water is added to the external phase it allows complete diffusion of acetone from internal phase to aqueous phase.
- This method is used for drugs and polymers which are soluble in polar solvents.

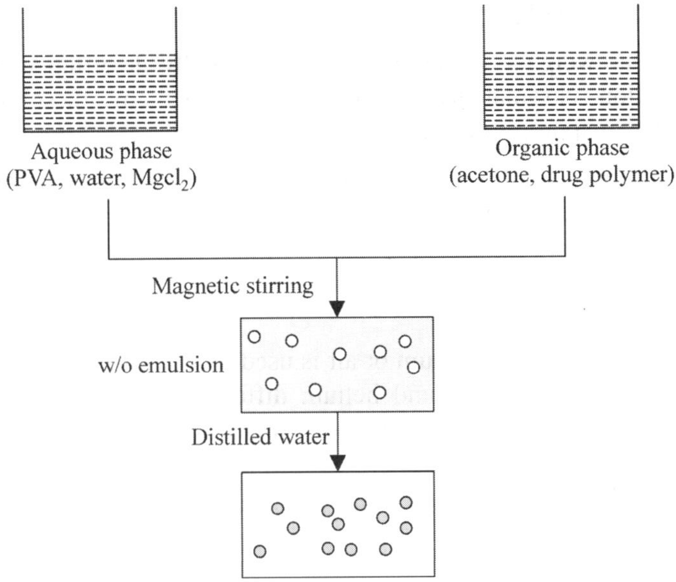

Fig. 1.6

Characterization/Evaluation of Nano Particles

1. Particle size and morphology
2. Molecular weight
3. Density
4. Surface properties
   (a) Surface charge
   (b) Surface hydrophobicity
5. Drug release and Drug loading

1. Particle Size and Morphology

One of the most important parameters of nanoparticles is particle size. It is determined by using following techniques,

(a) Photon correlation spectroscopy (PCS)

(b) Electron microscopy

  (i) Scanning electron microscopy (SEM)

  (ii) Transmission electron microscopy (TEM)

2. Molecular Weight

Molecular weights of nanoparticles are usually determined by Gel-permeation chromatography. It is determined by dissolving the particles in appropriate solvent.

Gel chromatography requires total dissolution of the polymer so molecular weight determination of cross linked polymers is not possible.

3. Density

Gas pycnometer with helium or air is used to measure the density. The value obtained with air and helium differs from each other due to specific surface area and porosity of structure.

4. Surface Properties

(a) Surface Charge

As surface charge determines the interaction of nanoparticles with biological environment and bioactive compounds, determining and studying the surface charge plays a important role.

The surface charge can be determined by

- Electrophoresis
- Laser Doppler Anemometry (LDA) (velocimeter)

(b) Surface Hydrophobicity

- Interaction of colloidal particles with the biological environment (e.g., protein adsorption, cell adhesion) are influenced by surface hydrophobicity.
- Biofate of nanoparticles and its contents are determined by hydrophobicity and hydrophilicity.
- Hydrophobicity and hydrophilicity of nanoparticles are determined by measuring angle of contact.

5. Drug Release and Drug Loading

Drug Loading

- High loading capacity is the property of the successful nanoparticulate system.

- Drug loading is done by two methods,
    - (i) In corporation at the time of production
    - (ii) Adsorption of drug after the formation of nanoparticles
- Drug loading and entrapment depend on the
    - Solid dissolution or solid dispersion
    - Molecular weight
    - Drug polymer interactions
    - Presence of end functional groups
    - Macro molecule or protein show good loading efficiency

Drug Release

- Drug release and polymer biodegradation are important consideration factors for a successful nanoparticulate system.

  Drug release rate depends on:
    - (i) Solubility of drug
    - (ii) Desorption of the surface bound/adsorbed drug
    - (iii) Drug diffusion through the nanoparticle matrix
    - (iv) Nanoparticle matrix erosion/degradation
    - (v) Combination of erosion/diffusion process
- Various methods which can be used to study the *in vitro* release of the drug are,
    - (i) Side-by-side diffusion cells with artificial or biological membranes
    - (ii) Dialysis bag diffusion technique
    - (iii) Reverse dialysis bag technique
    - (iv) Agitation followed by ultracentrifugation/centrifugation
    - (v) Ultra-filtration or centrifugal ultra-filtration techniques

Application of Nanoparticles

Nanoparticles are widely applied for different therapeutic application they are

- These are used in intracellular targeting of anti-infective to reduce the difficulty to treat intracellular infections in human body. E.g., ampicillin to nanoparticles for intra-bacterial infections.

- To reduce the toxicity and increase the therapeutic activity of cytostatic drugs.
- For specific targeting of anti-inflammatory drugs to minimize the side effects.
- Normal aqueous eye drops have half-life 1-3 minutes but nanoparticles has half-life of 15-20 minutes.
- In nuclear medicine they are used as carriers for radio nucleotides for diagnostic purpose.
- To improve the solubility and bioavailability of poorly soluble drugs.
- They are used as solid lipid nanoparticles in skin and hair care.
- To deliver drug across blood brain barrier.
- For targeted delivery of peptides and proteins.
- Used as adjuvant for vaccines. E.g., polymethyl methacrylate.

CHAPTER 2

SPECIFIC DRUG DELIVERY SYSTEMS FOR COLON

Introduction

Colon is the best site for absorption of protein and peptide drugs. These protein and peptide drugs are commonly administered by parenteral route which is expensive and inconvenient and if they administered by oral route these drugs are destroyed and made inactive in acidic environment of stomach and or by the enzymes of pancreatic juice in small intestine and so, when these drugs administered or targeted to colon, due to longer transit (20-30 hours), less peptidase activity, natural absorptive power and high response to absorptive power and high response to absorption enhancers made the colon a gunshot site for the delivery of systemic absorption of peptide and protein drugs.

- Colon drug delivery is also useful in development of pulsatile drug delivery systems. That is intentional delayed absorption of drug is required from the chronotherapeutic point of view in curing of diseases that have peak symptoms at definite periods. E.g., early morning symptoms are high in case of running nose and allergies. Night times asthma etc.
- Colonic drug delivery is found as a tool for improving systemic absorption of drugs like isosorbide mononitrate, metroprolol and theophylline.

Cross Linking in W/O Emulsion

- This method is widely used for nano encapsulation of drugs.
- This method involves emulsification of bovine serum albumin (BVA), human serum albumin (HSA) or protein aqueous solution in oil under high pressure homogenization.

Colonic Diseases

Before proceeding to the colon targeting, one must have an idea on the diseases that affects the colon.

Colonic diseases are broadly classified in to two types:

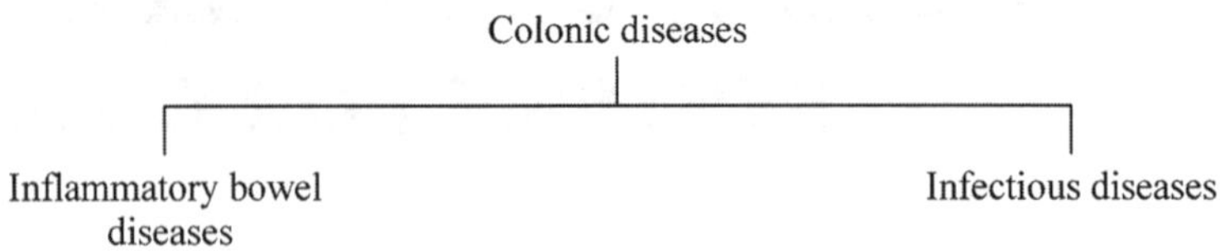

If the ulcerative colitis is neglected then it will develops into colon cancer.

Factors to be kept in mind for the designing of colon specific drug delivery system:

1. Physiological and anatomy of colon
2. pH of colon
3. Gastro intestinal transit
4. Micro flora in the colon

Physiology and Anatomy of Colon

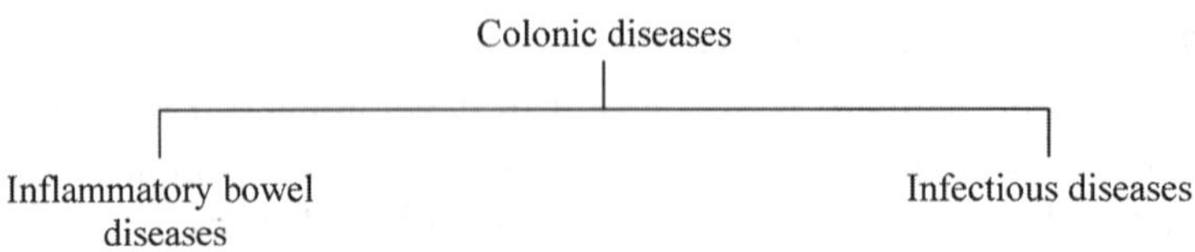

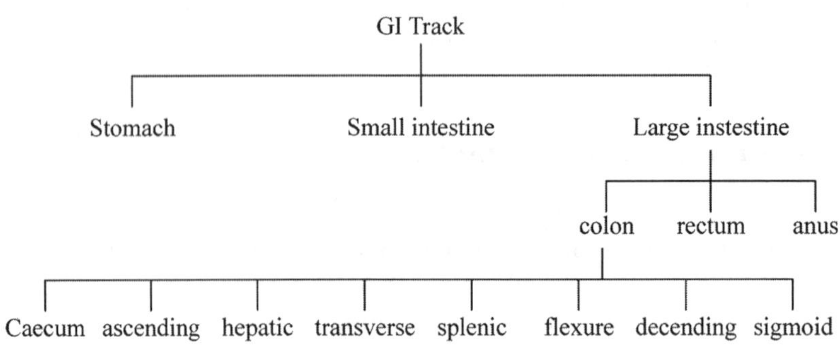

Colon is nearly 1.3-1.7 m long and the transverse colon is the longest part of all. The wall of colon is made up of four layers.

Fig. 2.1 Structure of colon.

The mucosa is subdivided into epithelium, lamina propria and muscularis mucosa. Within the lamina proporia of the blood vessels and lymphatic lacteals are located.

1. pH of the Colon

To determine the pH in the GI tract method used is radiotelemetry.

Part of the Colon	pH
Right Colon	6.4 ± 0.6
Mid Colon	6.6 ± 0.8
Left Colon	7.0 ± 0.7

Colonic pH has been shown reduced in disease states. This has been proved by Raimundo et al., the mean pH in a group of seven patients with untreated ulcerative colitis was found to be 4.7 ± 0.7 whereas patients under treatment has a pH of 5.5 ± 0.4.

2. Gastrointestinal Transit

Part of GIT	Transit Time (hr)
Stomach	<1(fasting) >3 & <12 (fed)
Small Intestine	3 – 4
Large Intestine	20 – 28

The arrival of an oral dosage form at the colon is dependent on the gastric emptying and the small intestinal transit time.

In fed conditions gastric transit will increase and the transit time in small intestine will remain constant and it does not depend upon fed and fasting conditions. Therefore in an average a dosage form would take a minimum of 4 hours to maximum of 14 hours to reach the colon and the colon transit time is 20-30 hours.

In disease conditions the colonic transit time changes as in diarrhea where the transit time increases and in constipation it will decrease.

3. Colonic Microflora

Aerobic and anaerobic bacteria are present on the entire GIT of human beings. The concentration of the colon is 10^{11} to 10^{12} CFU/ml. A large number of anaerobic bacteria are present in colon than that of aerobic bacteria. Examples of bacteria which are anaerobic include propiono bacterium, eubacterium, clostridium, bacteriodes and other sources.

The compounds taken orally are metabolized by the bacteria of the gut. The important metabolic reactions carried out by intestinal bacteria are as follows:

Metabolic Process	Type of Linkages
Hydrolysis	Glycosides, amides, esters
Reduction	C=C,Azo bonds, aldehydes, ketones
Dehydration	C & N
Dealkylation	N & O –alkyl groups

The metabolic actions of microflora are dependent on factors like age, intake of drugs and fermentation of the dietary residues, diseases in the GIT which can lead to either enhancement or inactivation of the drug compounds and sometimes increases the side effects of the drugs.

Drug Absorption in the Colon

For a drug to get absorbed into the colon it should have to be influenced by the following barriers or conditions.

(i) High viscosity of the colonic contents which makes difficult for a drug to diffuse from lumen to the site of absorption.

(ii) In lumen specific and non-specific drug binding increases the residence time of the drug and hence more enzymatic or environmental degradation occurs.

(iii) Binding of the drug to the mucus or drug mucus repulsion make the drug as a barrier in reaching the epithelial surface.

(iv) The most significant barrier in the colon is the epithelium and this is made of lipid bi-layer of individual colonocytes and the occluding junctional complex which extends between the colonocytes acts as a physical barrier for drug absorption.

(v) Drugs diffuse in the following ways in the epithelium.

  (a) Transcellular route (through the colonocytes).

  (b) Paracellular route (between adjacent colonocytes): Paracellular transport occurs most widely in the colon.

(vi) Local enzymatic degradation in the sub-mucosa of the colon can also act as a barrier for drug absorption.

In order to overcome these barriers absorbance enhancers are used to increase the drug absorption in the colon. These enhancers act mainly by three mechanisms which are as follows:

1. By destroying the integrity of lipid bi-layers.
2. By denaturation of membranes, they will alter the permeability of the drug into the colon.
3. By disrupting the intercellular occluding junctional complex and thereby increases the paracellular transport of the drug.

Examples of the absorption enhancers: (Muranishi et al.) Mixed micelles composed of either taurocholate or glycocholate with monoolein, oleic or lauric acid enhances the absorption of 5-fluoro uracil, heparin and bleomycin.

Approaches for Achieving Colon Specific Drug Delivery

This was developed by Kinget et al., Watts and Illum Ans van den Mooter et al.

1. pH dependant polymer coating.
2. Dosage form release based upon time.
3. Development of delivery system based on the metabolic activity of colonic bacteria.

1. pH-Dependant Polymer Coating

In this method the drug is first formulated as solid dosage form such as tablets, capsules and then coating them with pH sensitive polymers.

The polymer most widely used is methacrylic resins (Eudragits) which are available as water soluble and insoluble compounds.

Eudragit L --- water soluble at pH 6 or above.

Eudragit S --- water soluble at pH 7 or above.

5-Amino salicylic acid is available commercially coated with Eudragit-L (Mesazal®) and with Eudragit-S (Asacol®), which specifically releases the drug mostly in the colon.

The Eeudragit-S is made more specific polymer by partially methylating the free carboxylic acid groups.

$$\text{Eudragit 's'} \xrightarrow[\text{methylating}]{\text{partially}} \text{More specific}$$

Disadvantages

- Lack of consistency in dissolution of the polymer.
- Presence of bile acid residues, carbon dioxide, short chain fatty acids in the colon made the pH below 6 and hence this system will not release the drug at all.

2. Dosage forms Released based upon the Time

In this method the solid dosage form is mostly formulated in the form of capsule, half of this is non-disintegrating and the other half is enteric coated.

The enteric coated half dissolve on entering the small intestine and a hydrogel is placed as a cap to the non-disintegrating half and the gel swells (which depends upon cross-linking of the gel). After a predetermined time the gel swells mostly and comes out from the non-disintegrating half of the capsule and then the drug releases. And in this case the swelling of the hydrogel is pH independent and only depends on the swelling time.

Disadvantages

The site specificity of the timed release dosage form is very poor; the reason is because of large irregularities in gastric emptying times.

3. Development of Delivery System based upon the Metabolic Activity of the Colonic Bacteria

Mostly the colonic bacteria carries the metabolic reactions like hydrolysis and reduction. These actions are used for targeting the drug to the colon in the following ways:

(a) Coating with the Biodegradable AZO Polymers

The dosage forms are coated with the Azo-polymers most widely because of reduction of azo bonds is a general reaction of colonic bacteria specifically and that the dosage form after reaching the colon only the reduction of azo bond takes place and thereby we can achieve the site specific drug delivery.

The azo polymers which are having high degree of hydrophilicity are degraded by the colonic bacteria.

Saffron et al., developed co-polymers of styrene and 2-hydroxyethyl methacryalate which are cross linked with divinyl azo benzene and N,N'-bis (β-styrene sulphonyl)-4,4' diamino azo benzene to coat dosage forms of insulin and vasopressin, on entering the colon the coatings are degraded by bacterial azo reductases thereby releasing the drug.

Theophylline capsules coated with azo polymers based on 2-hydroxy ethyl methacrylate and methacrylic acid. The peak concentration of the drug was found higher when capsules were injected in the ceacum as compared to that of intestine.

Advantages

- Have more site specificity.
- Scope for administering large doses of drugs.

Disadvantages

- Azo polymers are expensive and has poor film forming capabilities.
- Azo polymers are not suitable for the delivery of the drugs with narrow therapeutic windows.

Dextrans

Dextrans are the macromolecules have been the field of interest in colon specific drug delivery. Drug is formulated as the dextran ester prodrug and on reaching this prodrug to the colon it is acted upon by the colonic-dextrases and thus results in the release of active drug.

$$\text{Dextran ester prodrug} \xrightarrow{\text{Colonic dextranases}} \text{Release of active drug}$$

Example: Dextran ester prodrug of naproxen was prepared by Larsen and evaluated for drug release and found that the drug release in the colon and the ceacum is 15-17 times faster than that of small intestinal drug release.

One important criterion for the colonic delivery of the dosage forms using the dextran ester prodrug method is that the drug has to possess a carboxy functional group.

Anti-inflammatory glucocorticoids do not possess carboxylic acid groups and thus they must be chemically transformed.

Example: Prednisone, prednisolone and dexamethasone were attached to the dextran using succinic acid as a spacer and thus introduced carboxylic acid group and the *in vivo* studies shows that the drug was rapidly released into the colon.

Disadvantages

- Prodrugs are expensive; they need the chemical modification of the drug.
- And many other substances like hydrogels; polysaccharaides like peptides etc., guar gum, inulin, chondrotin sulphate has also been used as colon drug targeting polymers.

Evaluation of Colon Specific Drug Delivery System

The colon specific drug delivery systems are evaluated by the following methods:

1. *In vitro* methods
2. *In vivo* methods

In vitro Methods

First the dosage form is observed assessed by conducting drug release studies in 0.1 N HCl for 2 hours (mean gastric emptying time). And in the pH 7.4 Sorensen's phosphate buffer for 3 hours (mean small intestinal transit time) using the USP dissolution apparatus (or) flow through dissolution apparatus.

During this period the dosage forms does not/show a little amount of drug release. And the ability of dosage form to release drug in colon is tested by incubating it in a buffer medium in presence of either enzymes of colon (e.g., pectinases, dextranases etc.) or by the rat, guinea pig colon/caecal contents.

The amounts of drug released at different intervals during study are determined. Another *in vitro* method is by placing the dosage form in a fermentor with the presence of human colonic bacteria in a suitable medium under anaerobic conditions and amount of drug released at different time intervals is found out.

In vivo Methods

Different techniques are available for testing the *in vivo* performance of the colon drug delivery system in human beings which is as follows:

1. Endoscopy

In this technique, the patient should have to swallow the endoscopic tube and a fibre scope (gastroscope) is used to monitor the behavior of the dosage form.

Demerits

This method requires administration of a mild sedative for the swallowing of the endoscopic tube, which will alter the GI motility.

2. String Technique

In this technique, the dosage form is attached to a string piece and the subject should swallow the dosage form leaving string hanging from mouth.

At different time periods the dosage form is withdrawn with the help of string and it is examined for the disinfection.

Demerits

As strip also passes through the GIT, it will alter motility and physical environment.

3. Roentgenography

In this method, a radio-opaque material is included into the dosage form and the dosage form is visualized by the use of x-rays

Example: By incorporating $BaSO_4$ into a dosage form we can able to follow the movement, location and the integrity of the dosage form after oral administration by placing the subject under a fluoroscope and talking a series of x-rays at various time periods.

Demerits

- Information cannot be obtained in continuous process.
- Use of x-rays involves subjects for exposing to high radiation.

4. Gamma Scintigraphy

This is the most widely used non-invasive technique for studying the *in vivo* behavior of solid oral dosage forms under normal physiogical conditions.

This method requires the presence of a gamma emotting radioactive isotope in the dosage form that can be detected *in vivo* by an extended gamma camera.

E.g., ^{99m}Tc, ^{111}In.

Advantages

- Gives very little radiation exposure.
- Gives both qualitative and quantitative results.
- Allows for *in vivo* evaluation of a dosage form under normal physiological conditions.

CHAPTER 3

GASTRO-RETENTIVE DRUG DELIVERY SYSTMES

Introduction

Oral administration is the most convenient and preferred means of any drug delivery to the system circulation. Oral controlled release drug delivery have recently been of increasing interest in pharmaceutical field to achieve improved therapeutic advantages, such as, ease of dosing administration, patient compliance and flexibility in formulation. Drugs that are easily absorbed from gastrointestinal tract (GIT) and have short half-life are eliminated quickly from the systematic circulation. Frequent dosing of these drugs is required to achieve suitable therapeutic activity. To avoid this limitation, the development of oral sustained-controlled release formulations is an attempt to release the drug slowly into the GIT and maintain an effective drug concentration in the systematic circulation for a long time. After oral administration, such a drug delivery would be retained in the stomach and release the drug in a controlled manner, so that the drug could be supplied continuously to its absorption sites in the GIT.

These drug delivery systems suffer from mainly two adversities i.e. the short gastric retention time (GRT) and unpredictable short gastric emptying time (GET), which can result in incomplete drug release from the dosage form in the absorption zone (stomach or upper part of small intestine) leading to diminished efficacy of administered dose. To formulate a site-specific orally administered controlled release dosage form, it is desirable to achieve a prolonged gastric residence time by the drug delivery. Prolonged gastric retention improves bioavailability, increases the duration of drug release, reduces drug waste, and improves the drug solubility that are less soluble in a high pH environment. Also prolonged GRT in the stomach could be advantageous for local action in the upper part of the small intestine, e.g. treatment of peptic ulcer, etc.

Gastro retentive drug delivery is an approach to prolong gastric residence time, thereby targeting site-specific drug release in the upper GIT for local or systemic effects. Gastro retentive dosage forms can remain in the gastric region for long periods and hence significantly prolong the GRT of drugs. Over the last few decades, several gastro retentive drug delivery approaches being designed and developed, including:

- High density (sinking) systems that is retained in the bottom of the stomach,
- Low density (floating) systems that causes buoyancy in gastric fluid,
- Mucoadhesive systems that causes bioadhesion to stomach mucosa.
- Swellable systems which limits emptying of the dosage forms through the pyloric sphincter of stomach,
- Super porous hydrogel systems,
- Magnetic systems etc.

Definition

Gastro retentive drug delivery systems are the novel formulation systems which are designed to prolong the gastric retention time and to have target-specific drug release in the upper part of stomach for local or systemic therapeutic effects.

Their gastric retention is prolonged either by retaining them in stomach for longer periods or by floating the drug in gastric fluid for a period of time to increase the disintegration time.

Advantages of Gastro Retentive Drug Delivery Systems

1. The bioavailability of those agents which get metabolized in the upper part of GIT can be enhanced rapidly.
2. For dosage forms with shorter half-life this approach causes a decrease in dosing frequency and provides good patient compliance.
3. This drug delivery system is also used to overcome the adversities caused due to gastric retention and gastric emptying time.
4. Due to their prolonged release of drugs they are useful in treatment of disorders related to stomach and small intestine.
5. This drug delivery system can be used for drugs with narrow therapeutic index because some adverse effects are concentration dependent.

6. This drug delivery system minimizes the fluctuation of drug concentration and there by obtaining improved selectivity in receptor activation.
7. The sustained release property of this drug delivery system helps to improve the pharmacological and chemical outcomes of the active drug.

Disadvantages of Gastro Retentive Drug Delivery Systems

1. There are situations where gastric retention is not desirable.

   E.g., aspirin causes gastric lesions in the stomach if present for longer periods.
2. Drugs that are unstable in acidic environment should not be formulated for this type of system.
3. Drugs that are easily absorbed through GIT cannot be formulated as the absorption remains same in both cases. E.g., Isosorbide dinitrate.

Anatomy and Physiology of Gastrointestinal Tract

The GIT can be basically divided into pharynx, esophagus, stomach, small intestine and large intestine. Small intestine further consists of duodenum, jejunum, and ileum. Gastric emptying generally occurs during both fasting as well as fed states. During fasting state an interdigestive series of events take place which cycle both through stomach and intestine. This has the following phases,

- Phase 1 (basal phase) lasts from 40 to 60 minutes with rare contractions.
- Phase 2 (preburst phase) lasts for 40 to 60 minutes with intermittent action potentials and contractions.
- Phase 3 (burst phase) lasts for 4 to 6 minutes. It includes intense and regular contractions for short period.
- Phase 4 lasts for 0 to 5 minutes and occur between phases 3 and 1 of two consecutive cycles.

Drug Candidates for Gastro Retentive Drug Delivery Systems

- Drugs those are locally active in stomach.

  E.g., Misoprostol, antacids.
- Drugs with narrow absorption window in GIT.

  E.g., Furosemide, riboflavin.

- Drugs those are unstable in intestinal or colonic fluids.
  E.g., Captopril, metronidazole.
- Drugs that kill the microbial flora in intestine.
  E.g., Antibiotics.
- Drugs that has low solubility at high P_H
  E.g., Diazepam, verapamil.

Drugs for which Gastro Retentive Drug Delivery Systems is not Suitable

- Drugs with very limited acid solubility.
  E.g., Phenytoin.
- Drugs those are instable in gastric environment.
- Drugs that are intended for release in lower parts of GIT i.e. colon, rectum.
  E.g., 5-amino salicylic acid, corticosteroids.

Factors Affecting Gastric Retention of Dosage Forms

The important factors controlling the gastric retention time are as follows

1. Density of dosage form
2. Shape and size of dosage form
3. Nature of food taken
4. Miscellaneous

1. Density of dosage form

The density of dosage form has great influence in case of gastric retentive dosage forms as it determines the location of the dosage form and its therapeutic effect. Formulation with a density higher than gastric juice sinks and that with lesser density floats on the gastric fluid. A density of < 1.0 g/cm^3 is required to exhibit floating property of the formulation.

2. Shape and size of dosage form

The size and shape of dosage form has important influence on the retention time. The retention time for larger dosage form is greater than smaller ones because larger size will not allow it pass quickly through the pyloric antrum into intestine. Generally a diameter range of 7.5-9.9 mm is preferred than other forms. Ring and tetrahedron shaped formulations have better residence time than other shapes.

3. Nature of food taken

Nature of food intake, its viscosity, volume, calorie value and frequency of feeding have a profound effect on gastric retention of dosage form. The presence of food in stomach and the acidic nature of food generally cause an increase in the gastric retention time of the dosage form and increase its absorption rate.

4. Miscellaneous

Gender and age of patients has influence on the gastric emptying. Generally females have slower gastric emptying and an elderly person has a slower gastric emptying.

Classification of Gastro Retentive Drug Delivery Systems (GRDDS)

A. Floating DDS
B. Muco/Bio adhesive DDS
C. High density systems
D. Ion exchange DDS
E. Magnetic systems

A. Floating drug delivery systems

1. Effervescent systems
    - I. Gas generating systems
        - (a) Intra gastric single layer floating tablets
        - (b) Intra gastric bilayer floating tablets
        - (c) Multiple unit type floating pills
    - II. Volatile liquid/vacuum containing systems
        - (a) Intra gastric gastrointestinal DDS
        - (b) Inflatable gastrointestinal DDS
        - (c) Intra gastric osmotically controlled DDS

2. Non effervescent systems
    I. Single layer floating tablets
    II. Bilayer floating tablets
    III. Alginate beads
    IV. Hollow microspheres

1. Effervescent systems

I. Gas generating system

These are the systems which float in the gastric juice due to the release of gas in the formulation which makes them to float in the gastric juice.

(a) ***Intra gastric single layer floating tablets***

This type of drug delivery systems is also called as hydrodynamically balanced systems. These are generally formulated with CO_2 generating agents in the matrix tablet. These have a bulk density less than gastric fluid resulting in their floating of this DDS.

(b) ***Intra gastric bilayered floating tablets***

These are also compressed tablets with 2 layers i.e. immediate release layer and sustained release layer.

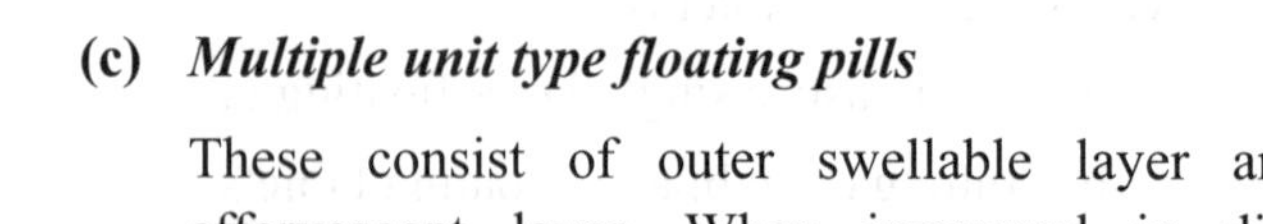

Intra gastric bilayered floating tablets

(c) ***Multiple unit type floating pills***

These consist of outer swellable layer and inner effervescent layer. When immersed in dissolution medium it first sinks and then forms swollen pills which float due to their low density.

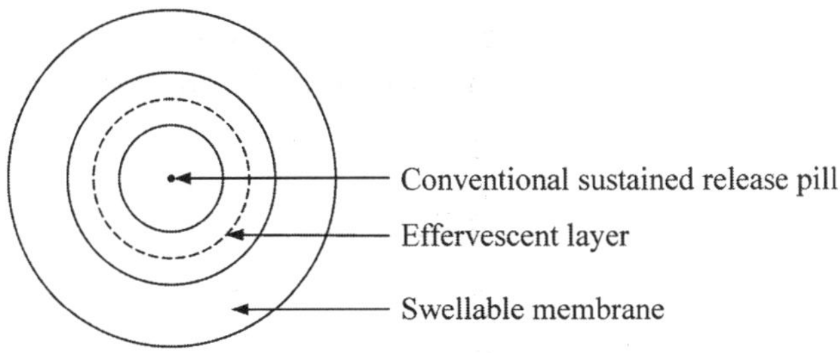

II. Volatile liquid/vacuum containing systems

(a) ***Intra gastric floating gastrointestinal drug delivery system***

In this the system is made to float in stomach because of the flotation chamber filled with vacuum or inert gas. The drug is encapsulated in a micro porous compartment.

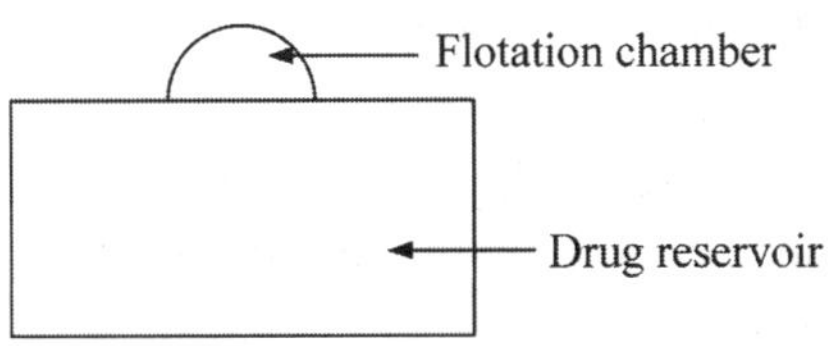

(b) ***Inflatable gastrointestinal delivery systems***

This consists of an inflatable chamber composed of liquid ether that gasifies at body temperature. After oral administration the capsule dissolves to release the drug in the mean time the chamber inflates and floats in the gastric juice leading to controlled release of the drug.

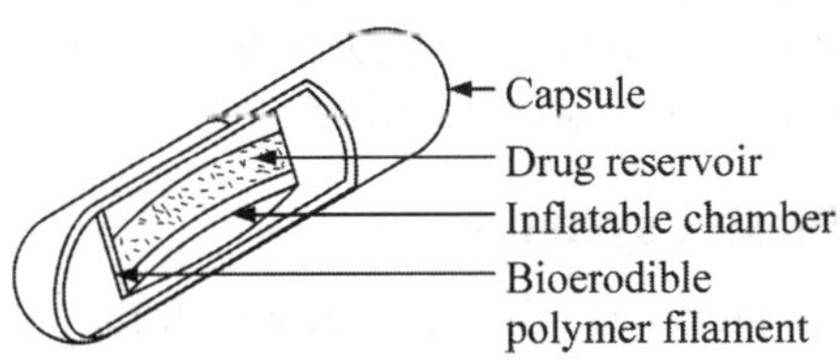

(c) ***Intragastric osmotically controlled drug delivery system***

These consist of a drug reservoir compartment and an osmotically active compartment. This osmotic compartment is enclosed by salt which when contacted with gastric fluid exerts osmotic pressure and causes the drug to release from the compartment.

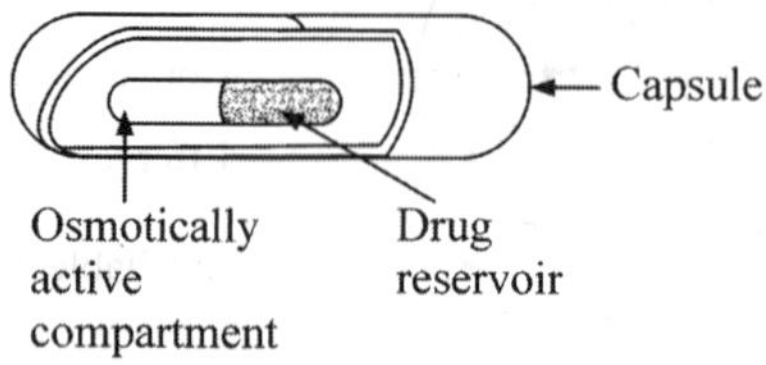

2. Non effervescent systems

I. Single layer floating tablets

These contain a gel forming hydrocolloid, which swells in contact with gastric fluid and maintains a bulk density less than unity. The trapped air in the polymers causes the buoyancy of the tablets.

II. Bilayer floating tablets

This has an immediate release layer for initial dose and a sustained release layer that release drug after forming a colloidal gel layer with gastric juice and remains buoyant in the stomach.

III. Alginate beads

These are prepared by immersing sodium alginate solution into calcium chloride aqueous solution, causing precipitation of calcium alginate beads with buoyant properties.

IV. Hollow microspheres

These are prepared by simple solvent evaporation/solvent diffusion methods. These DDS has the advantage of multiple unit system and good floating.

B. Muco/Bio adhesive systems

Mucoadhesive systems are the ones in which they will bind to gastric epithelial cell surface/mucin causing the drug to retain in the gastro intestinal tract. This phenomenon is based on the following mechanisms:

- Wetting theory, this deals with the ability of polymers to develop intimate contact with the mucous layer.
- Diffusion theory, proposes the interpenetration of mucin strands into the porous structure of polymer substrate.

- Absorption theory, states that adhesion is due to Vanderwaal's forces and hydrogen bonding.
- Electron theory, states an attractive electrostatic forces between the glycoprotein mucin and the bio adhesive material.

The polymers generally used in this for bioadhesion are polyacrylic acid, chitosan, cholestyramine, sodium alginate, polyethylene glycol, dextrin, etc.

C. High Density Systems

These systems are generally having a high density than gastric juice which causes them to sink to bottom of stomach i.e. rugae of stomach. These generally contain a density of about 3 g/cm^3. A range of 2.6-2.8 g/cm^3 is given as the threshold value for these systems. These are generally formulated by using high density inert coating material such as barium sulphate, zinc oxide, titanium dioxide, iron powder etc.

D. Ion Exchange DDS

These generally consist of DDS loaded with bicarbonate and negatively charged drug bound to resin. The formed beads are then encapsulated in a semi-permeable membrane to overcome the loss of CO_2. When this comes in contact with the gastric fluid causes the release of CO_2 and gets entrapped in the membrane which causes the DDS to float in gastric juice.

E. Magnetic systems

This gastric retention time mechanism is based on the principle that DDS contains a small internal magnet, and a magnet placed on the abdomen over the stomach. This has to be done with accuracy and precision to place the external magnet with improved patient compliance.

Drugs used in Formulation of Gastro Retentive Dosage Forms

Dosage form	Drugs
Floating tablets	Amoxicillin trihydrate, acetaminophen, ampicillin, captopril, ciprofloxacin
Floating capsules	Diazepam, furosemide, propranolol, misoprostol
Floating microspheres	Aspirin,griseofulvin,ibuprofen, terfenadine,
Floating granules	diclofenac sodium, indomethacin, prednisolone

Floating Drug Delivery Systems

Floating drug delivery is important in approaching gastric retention to obtain sufficient drug bioavailability. These are desirable for drugs with absorption in stomach. These have a bulk density less than gastric fluids and remains buoyant in the stomach for prolonged period of time. The requirements for floating drug delivery systems are,

- Contents should be released slowly and act as a reservoir for drug.
- Specific gravity lower than gastric contents (1.004-1.01 g/cm^3).
- Must form a cohesive gel barrier.

The low density of these formulations is attributed to the entrapment of air or due to incorporation of low density materials (fatty materials, oils, foam powder). Recent development is the single-unit floating system was proposed consisting of polypropylene foam powder, matrix forming polymers, drug and filler.

These are classified as

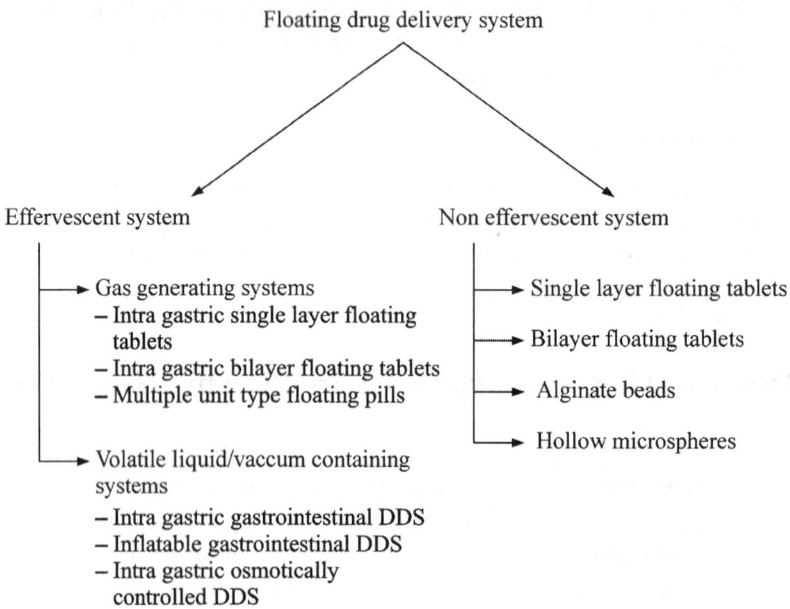

Formulation of Floating Drug Delivery Systems

The formulation generally include

(i) Active Pharmaceutical Ingredient (API)

(ii) Hydrocolloids

(iii) Inert fatty material

(iv) Rate release accelerants

(v) Rate release retardants

(vi) Buoyancy increasing agents

(i) Active pharmaceutical ingredient (API)

The API should be stable at the gastric environment, if not it is modified to achieve stability in the gastric environment. The drugs that have therapeutic effect in the stomach are generally preferred for this type of drug delivery system.

(ii) Hydrocolloids

These are generally added in a concentration of 20-75%. Hydrocolloids can be synthetic, anionic, and non-ionic like hydrophilic gums, modified cellulose derivatives.

E.g., Acacia, pectin, chitosan, agar, casein, bentonite, etc.

(iii) Inert fatty materials

5-7% of inert fatty materials can be added to the formulation. These should have a specific gravity less than 1 and can be used to decrease the hydrophilic property of formulation which increases the buoyancy of the formulation.

E.g., Bees wax, long chain fatty alcohol.

(iv) Rate release accelerants

These in the range of 5-60% can be added to increase the rate of drug release from the polymer matrix of the formulation.

E.g., Lactose, mannitol.

(v) Rate release retardants

These in the range of 5-60% can be added to retard the rate of drug release from the polymer matrix of the formulation. These causes increased bioavilability of drug.

E.g., Dicalcium phosphate, talc, magnesium stearate.

(vi) Buoyancy increasing agents

These agents help in enhancing the buoyancy of formulation and keep them floating them on the gastric fluids. E.g., Ethyl cellulose.

This formulation is used to design the dosage form of the drug and can be given as,

- Tablets
- Capsules
- Microspheres
- Granules

Evaluation of Floating Drug Delivery Systems

The evaluation of these drug delivery systems involves the following tests

1. Measurement of buoyancy

The measurement of buoyancy of the dosage form is determined both in deionised and stimulated meal, to determine the difference. These results determined that the high molecular weight polymers with slow rate of hydration have enhanced buoyancy in stimulated meal medium than that in deionised water.

2. Floating time

The time required for the dosage form to float is called floating lag time. The time for which the dosage form floats is termed as floating time or flotation time. This is generally determined by using the USP Dissolution apparatus using stimulated gastric meal or 0.1 mole lit^{-1} hydrochloric acid solution maintained at 37°C. A solution of 900 ml of 0.1 mole lit^{-1} HCl is taken in the beaker and both the floating lag time and flotation time are determined.

Gohel proposed a more relevant model to determine the floating time which mimic the gastric fliuid. His model consists of a beaker of 100 ml capacity with a side tube and a burette at the top to ingest gastric juice at the rate of 2 ml/min. This overcomes the problem of sticking the dosage form to the paddle in USP Dissolution apparatus. In this the drug release follows the zero order kinetics and showed good *in vitro-in vivo* correlation of drug release.

3. Dissolution

Dissolution test is performed by USP Dissolution apparatus and estimated by periodically collecting the samples from the medium and replacing them. The collected samples are analyzed for the concentration of the drug in the samples.

1. In case if formulated as tablets then content uniformity, hardness and friability tests are done to evaluate them.
2. Drug loading and drug entrapment efficiency, particle size analysis and surface characterization; (For floating microspheres and beads).
   (a) Drug loading is determined by taking an accurate quantity of dosage and crushing them in the mortar and dissolving the dosage in suitable dissolution medium and analyzing the liquid to determine the concentration of the solution and to determine the drug loading. It is calculated by dividing the amount of drug in the sample by weight of the total beads or the microspheres.
   (b) Particle size analysis is generally done in dry state using optical microscopy method.
   (c) The external and cross sectional morphology is determined by Scanning Electron Microscopy.

4. X-ray/Gamma scintigraphy

This is a very popular evaluation parameter for floating dosage form now a day. It helps to locate dosage form in the GIT and by which gastric emptying time and the passage of dosage form in the GIT can be predicted and correlated. In this method the inclusion of a radio-opaque material into a solid dosage form enables it to be visualized by x-rays. Similarly, the inclusion of a γ-emitting radionuclide in a formulation allows indirect external observation using a γ-camera or scintiscanner. In case of γ-scintigraphy, the γ-rays emitted by the radionuclide are focused on a camera, which helps to monitor the location of the dosage form in the GIT.

5. Pharmacokinetic studies

Pharmacokinetic studies are the integral part of the *in vivo* studies. Sawicki studied the pharmacokinetics of verapamil, from the floating pellets containing drug, filled into a capsule, and compared with the conventional verapamil tablets. The t_{max} and AUC values for floating pellets were comparatively higher than those obtained for the conventional verapamil tablets. There is no difference between the C_{max} values of both the formulations, suggesting the improved bioavailability of the floating pellets compared to the conventional tablets. An improvement in bioavailability has also been observed with piroxicam in hollow polycarbonate microspheres administered in rabbits. The microspheres showed about 1.4 times more bioavailability, and the elimination half-life was increased by about three times than the free drug.

The currently available polymer-mediated non-effervescent and effervescent FDDS, designed on the basis of delayed gastric emptying and buoyancy principles, appear to be a very much effective approach to the modulation of controlled oral drug delivery. It is suggested that future research work in the FDDS should be aimed at discovering means to control accurately the drug input rate into the GI tract for the optimization of the pharmacokinetic and toxicological profiles of medicinal agents.

CHAPTER 4

LIPOSOMES

Liposome: (meaning: lipid bodies)

- Defined as vesicle of lipid bilayer which encloses an aqueous compartment.
- Phospholipids are most commonly used lipid. Sphingolipids, glycolipids and steroids are also been to prepare liposome. Their size ranges from 25-500 nm.

Classification

Depending upon their structure liposomes are classified as follows:

1. Unilamellar vesicle (ULV)
2. Multilamellar vesicle (MLV)
3. Small unilamellar vesicle (SULV)
4. Large unilamellar vesicle (LULV)
5. Oligolamellar vesicle (OLV)

1. Unilamellar vesicle

These are made of single bi-layer of lipids.

- small unilamellar vesicle of size 20-40 nm
- medium unilamellar vesicle of size 40-80 nm
- large unilamellar vesicle of size 100-1000 nm
- giant unilamellar vesicle of size more than 1000 nm

2. Multilamellar vesicle

These are made of concentric bi-layers of lipid enclosing a small internal volume.

- Size 100 nm -1 micron and consist of 5 or more lamella.

3. Oligolamellar vesicle

These are made of 2-10 bi-layer of lipid membrane surrounding large internal volume.

- Liposome includes a large variety of drug (anti-neoplastic, antibiotics), peptides/proteins (including antibiotics), viruses and bacteria.
- Water soluble drugs are trapped in aqueous compartment while lipophilic once are incorporated in lipid phase of liposome.
- At controlled rate Intra muscular and subcutaneous injected liposome delivers drug, while intravenous administration selectively target them into reticuloendothelial cells and phagocytic cells.

Problems with Liposomes

1. First liposomes are most important part, unable to leave the general circulation and are unlikely to reach most cell types. Unencapsulated or free drug typically diffuse through capillary wall into tissues but liposomes are too large to pass through the wall of capillaries in most organs.
2. Body immune system turned out somehow to recognize the liposome and remove them from circulation.
3. Liposomes somehow enter and accumulate in extracellular spaces of reticulo endothelial cell, and organs probably prevent liposome from circulation long enough to reach many targeted cells and tissue efficiently.
4. Liposomal drug are effective against disease of reticulo endothelial cells
5. Liposome can improve the effectiveness and reduce toxicity of drug administration for such diseases as systemic fungal infection, leishmaniasis and cancer.

Applications

1. Liposome in anticancer therapy

Liposome is used for administering anticancer drug owing to toxicity due to site specific targeting, increasing half-life and increasing stability of drug.

E.g., Malignant tumours and leukaemia.

(a) Doxorubicin: Acts rapidly dividing cells is used in treatment of tumours and leukaemia.

Side effect of doxorubicin:

1. Progressive irreversible damage to heart.
2. It also attack hair follicle cell.
3. Act on intestinal cell and cell of immune system.

Causes: Hair loss, nausea, vomiting and immune suppression.

Liposomal doxorubicin is as effective as free doxorubicin but is less toxic to heart.

2. **Liposome as a carrier of vaccine**

Liposome may be vehicle for producing safe and potent antiviral, antibacterial and antiparasite vaccine.

3. **Formulation aid**

Hydrophilic drug such as cyclosporine is usually formulated in surfactants and organic co-solvents for administering in humans. These solubilizers may cause toxicity. In contrast liposomes are made up of lipid which are relatively non-toxic, non-immunogenic, biocompatible and biodegradable molecule, and can encapsulate a broad range of water insoluble drug.

4. **Liposomes are used in the treatment of systemic fungal infection**

Liposomal amphotericin used systemic infection more effectively than free drug. Since more liposomal doxorubicin could be given without increasing toxicity.

E.g., Amphotericin B, an antifungal is highly toxic to kidney.

5. **It is used in treatment of enzyme deficiency diseases**

Liposome has been advantageously used for delivery of enzyme into cell liposome. The enzymes are efficiently shielded in carrier. They have high degree of thermal stability and & protected against inactivation by proteolysis or by antibodies.

6. **Used in treatment of leishmaniasis**
7. **Liposomal formulations have been successfully used in treatment of rheumatoid arthritis.**

8. **Liposome can be used as carrier**

   Liposome can be used as carrier for both water and lipid soluble agents. Substances entrapped in liposome include enzyme, glycolipids, biological response modifiers, and antifungal, chelating agents, markers, vaccines and chromosomes.

9. **Metal toxicity**

   liposome EDTA is used in heavy metal poisoning. The encapsulated drug was far more effective than free drug in enhancing the removal of plutonium from tissue and in promoting.

10. **Involved in extracellular drug delivery and sustained release drug delivery.**

Advantages

1. Direction

   Liposome can target a drug to the intended site of action in body, thus enhancing its therapeutic efficacy.

   - Liposome may also direct a drug away from those body sites that are particularly sensitive to toxic action of it.

2. Duration

   Liposome act as depot from which entrapped compound is slowly released over time.

   - Such a sustained released process can be exploited to maintain therapeutic drug level in blood stream or at local administration site for prolonged period of time.

3. Protection

   Drug incorporated in liposome, in particular those entrapped in the aqueous interior are protected against the action of determinantal factors (e.g., Degradative enzyme present in host).

4. Internalization

   Liposome can interact with target cell in various ways and are therefore able to promote the intracellular delivery of drug more than in the free form (i.e. non capsulated) would not be able to enter the cellular interior.

Preparation of Liposomes

All the methods of preparing liposome involve 3 or 4 basic stages:

- Mixing of compounds of liposome in organic solvents.
- Drying down the lipids from organic solvent.
- Dispersion of liquid in aqueous media.
- Purification of resultant liposome.
- Analysis of final product.

1. The basic starting point of preparation of well defined liposome is to use well characterized lipid. Unsaturated lipids are normally used.

   E.g., Phosphotidyl choline and saturated lipid such as DMPC.

2. First the bi-layer forming elements are mixed in volatile organic solvents or solvent mixture such as chloroform, ether, ethanol or combination of these. The predominant bi-layer formation element is usually a neutral phospholipid such as cholesterol, is often included to provide great stability. A charge species may be added (5-20%) to prevent aggregation.
3. Small amounts of antioxidants are also included when poly unsaturated natural lipids are used.
4. Ones a suitable solution of lipid components is made, the mixture may be filtered to remove minor insoluble components or ultra-filter to lower pyrogens.
5. The solvent subsequently removed under conditions (pressure and temperature) that ensure no phase separation of components of mixture take place.
6. Following solvent removal the dry lipid mixture is usually hydrated by exposure to an aqueous medium containing dissolved solutes including buffer, salts and chelating agents.
7. Hydration more than any other step influence type of liposome formed (number of layers, size and entrapped volume). The nature of dry lipid, its surface area and its porosity (thin to thick film to fine powder, granular pellets etc.) is of particular important. The condition under which the dry lipid first encounters aqueous phase is critical and must be carefully controlled.

8. The uncapsulated drug can be removed by variety of techniques such as centrifugation, dialysis, difilteration and recovery.
9. After dispersion in an aqueous medium liposomal suspension is formed and then it is dried in the vacuum dryer and kept at particular pH buffer and temperature.
10. Purification
    - It is done by column chromatography.
    - TLC/HPTLC using sephadex 25 as stationary phase or by dialysis process using semi permeable membrane.
11. Analysis of final product
    - Done by NMR, X-RAY, Ultracentrifugation, light scattering.

Preparation of MLV

- It is simple process and it has been claimed that they form spontaneously.
- MLV's are suitable for encapsulation of variety of substances including drug.
- Drawback: Low encapsulation in terms of aqueous space per mole of lipid (0.2-1.0 micron).

Preparation of ULV

- Sonication of dispersion of phospholipid results in formation of microvesicle of 500Ådiameter.
- Drawback: Low encapsulation efficiency of aqueous space range of 0.1-1 micron.

Following are the various methods of preparation of Liposome:

1. **Detergent dialysis method**

   The principle of liposome preparation is the controlled and reproducible detergent removal from equilibrium solutions of mixed lipid/detergent micelles at definite temperature resulting in formation of small ULV'S.

   Method employed to remove the detergent from mixture are centrifugation, gel filtration and fast controlled dialysis. Rate of detergent removal as well as detergent to lipid ratio have been shown to play role in determining the size of liposome. Detergent with high CMC is removed by dialysis much more efficiency than low CMC.

2. **Solvent dispersion method**

    1. ***Reverse phase evaporation:*** Increase in ratio of aqueous space to lipid and can encapsulate large percentage of initial aqueous phase are advantages.
        - In this, added the aqueous material directly to the phospholipid solvent mixture and sonicated, as a result forms the stable water in oil emulsion. The organic solvent is then removed by rota evaporation drug solvent is removed.
        - A lipid paste is formed which upon further solvent evaporation takes place which form liposome suspension.
        - Liposome preparation is dependent upon formation of stable emulsion.
        - Disadvantage: Non encapsulated macromolecule such as protein which get denatured by organic solvent.
    2. ***Ethanol injection method:*** In ethanol lipids are dissolved and rapidly injected into a buffer solution where they spontaneously form small unilamellar vesicles.
        - This procedure is simple, gentle and rapid.
        - Disadvantage: Dilute preparation of liposome with low encapsulation efficiency is obtained.

3. **Solvent vaporization method**

    Dissolved the lipid in organic solvent and placed in a syringe. Inject the lipid solution at controlled rate into aqueous solution at temperature which causes the organic solvent to evaporate. The resultant suspension can be dialyzed or passed through gel filtration column to remove the unentrapped solute and residual organic solvent.

4. **Dehydration-rehydration technique:**

    This technique is recently developed. Method offer high entrapment of drug. Don't require sonification, detergents or organic solvents.

    - Drug solution is mixed with suspension of empty liposome and mixture is freezed and dried.
    - Upon rehydration liposome having large amount of drug is obtained.

- Unentrapped drug is removed by a process of technique dialysis.
- Preparation contains only entrapped drug is freezed dried and liposomal integrity can be maintained.

Evaluation of Liposomes

Liposomal formulation after their formulation and processing for a specified purpose are characterized to ensure their predictable *in vitro* and *in vivo* performance. The liposome prepared from different methods may have different characteristics and these differences do have an impact on their behavior *in vivo* and *in vivo*.

Three different types of evaluation are performed to check the efficiency of the prepared liposome. They are evaluation of

(a) **Physical parameters:** Physical parameters includes size, shape, surface features, phase behavior and drug release profile.

(b) **Chemical parameters:** Chemical parameters include those studies which establish the purity and potency of various liposomal constituents.

(c) **Biological parameters:** Biological parameters are helpful in establishing the safety and suitability of the formulations for the *in vivo* use or for the therapeutic applications.

1. Determination of particle size and shape

Vesicle size and shape can be assessed using various electron microscopic techniques.

(a) **Transmission electron microscopy:** It is useful for the observation of morphology of the liposome. In this technique samples were placed on to copper grill covered with nitrocellulose. They were dried at room temperature and then were examined using a Transmission Electron Microscopy (TEM).

(b) **Atomic force microscopy (AFM):** The shape and surface morphology of the liposome can be determined by using atomic force microscope. The dried liposome were suspended in distilled water and sonicated before measurement to obtain a homogenous suspension. A drop of the suspension was mounted on the metal slabs, air dried and was scanned by the AFM.

2. Determination of surface charge

The positive, negative or neutral charge on the surface of liposome is due to the composition of the head group. The surface charge of liposome governs the kinetic and extent of distribution *in vivo* as well as interaction with the target cell.

In general two methods are employed to assess the charge namely,

(a) Free flow electrophoresis

(b) Zeta potential measurement

From the mobility of the liposomal dispersion in a suitable buffer, the surface charge on the vesicles can be calculated.

Zeta potential determination was based on electrophoretic mobility of the nanoparticles in dilute aqueous suspension.

3. Encapsulation efficiency and trapped volume

Encapsulation efficiency and trapped volume determine the amount and rate of entrapment of water soluble agents in the aqueous compartment of liposome.

(a) Encapsulation efficiency

The encapsulation efficiency describes the percentage of the aqueous phase and hence the percentage of water soluble drugs that becomes ultimately entrapped during preparation of liposome and is usually expressed as percent entrapment/mg lipid. Encapsulation efficiency is assayed using two techniques including minimum centrifugation method and protamine aggregation method.

(b) Mini column centrifugation

It is generally used both as a means of purification and separation of liposome on a small scale and analysis of a liposomal dispersion to determine encapsulation efficacy.

(c) Protamine aggregation method

This method is used for neutral and negatively charged liposome. Liposomal dispersion can be precipitated with a protamine solution and subsequently centrifugation at 2000 rpm. By analyzer the material in the supernatant in the liposome pellet the encapsulation efficiency of the entrapped material can be estimated.

4. Drug release rate

The rate of drug release from the liposome can be determined by *in vitro* assays which help to predict the pharmacokinetic and bioavailability of the drug. However quantitative *in vivo* studies are found to be more complete.

5. Determination of phospholipid

The phospholipid content of liposome can be determined by two indirect assays. They are

(a) **Bartlett Assay**: This method of determining the phospholipid is very sensitive and may produce erroneous results in the presence of even trace amount of inorganic phosphate. Therefore borosilicate glass tubes and double distilled water is used.

(b) **Steward Assay**: This assay overcomes the drawback of Bartlett assay, but can not be used to evaluate those samples which contain the mixture of unknown phospholipid.

6. Cholesterol analysis

Quantitative analysis performed using a capillary column filled with fused silica.

7. Quantitative analysis

In this technique, sample is reacted with a reagent and the absorbance of purple color complex is measured at 610 nm.

Liposome offers several advantages for controlled drug delivery which involved in treatment in anticancer therapy, systemic fungal infection, enzyme deficiency diseases and also used as formulation aid.

CHAPTER 5

MICROSPHERES

Definition

Micro indicates particle size from 1 μm to 1000 μm, spheres means spherical particles. Microspheres are defined as small spherical particles, with diameter ranges from 1 μm to 1000 μm (1 mm). It is a monolithic spherical structure with drugs or active principle ingredient distributed throughout the matrix either as molecular dispersion or as dispersion of particles. Sometimes they are also called as microparticles, microbeads and beads.

Types of Microspheres

1. Glass microspheres
2. Polymer microspheres
3. Ceramic microspheres
4. Solid and Hollow microspheres

1. Glass Microspheres

- These are mostly used as fillers and volumizers for weight reduction.
- As retro-reflector for highway safety.
- As additive for cosmetics.

2. Polyethylene microspheres

- Polyethylene microspheres are commonly used as permanent or temporary filler.
- These are creating porous structures in ceramics and other materials due to their lower melting temperature.

- These are highly desirable for flow visualization, fluid flow analysis, microscopy techniques; health sciences process trouble shooting etc. due to the high sphericity of colored and fluorescent polyethylene microspheres.
- Charged polyethylene microspheres are used for electronic paper digital displays.

3. Ceramic microspheres

- Mostly used as grinding media.

4. Solid or hollow microspheres

These are potential approach for gastric retention. They worked on the principal mechanism of floatation to achieve gastric retention.

Commonly used Materials for Preparation of Microspheres

1. Non-biodegradable materials:
   (a) Epoxy polymers
   (b) Acrolein
   (c) Glycidylmethacrylate
2. Biodegradable materials:
   (a) Polyalkylcyanoacrylates
   (b) Lactides, glycolides and their co-polymers
   (c) Polyanhydrides
3. Natural materials:
   (a) Carbohydrates: Starch, Aga rose, Chitosan
   (b) Proteins: Albumin, Gelatin, Collagen
4. Chemically modified carbohydrates:
   (a) Poly(acryl) starch
   (b) Poly(acryl)dextran

Methods of Preparation

Several methods are available for the preparation of microspheres. Those are,

1. Spray drying
2. Solvent evaporation

3. Multiple emulsion method
4. Precipitation-chemical cross linking
5. Thermal cross linking
6. Ionotropic gelation
7. Complex coacervation
8. Wet inversion

1. Spray Drying

In this method, first the polymer is dissolved in a suitable volatile organic solvent such as dichloromethane, acetone etc. and the drug is dispersed in the polymer solution by high speed homogenization. Then the dispersion is atomized in a stream of hot air and it leads to the formation of small droplets. The microspheres are formed due to instantaneous evaporation of solvent from the small droplets. The formed micro particles are separated from the hot air by means of cyclone separator and the traces of solvent is removed by means of vacuum drying.

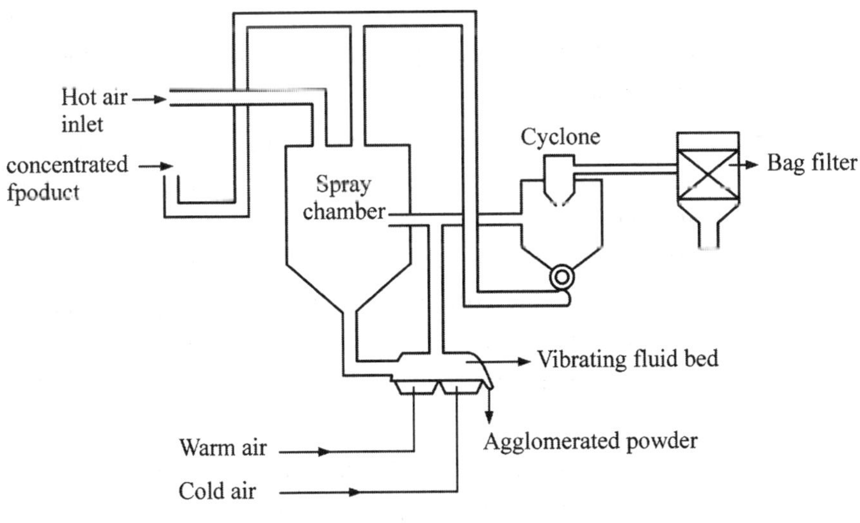

Fig. 5.1 Preparation of Microspheres by Spray drying.

Advantages

- This method gives microspheres in a size range of 1-100 μm.
- This process is rapid and it leads to the formation of porous microspheres.

2. Solvent Evaporation

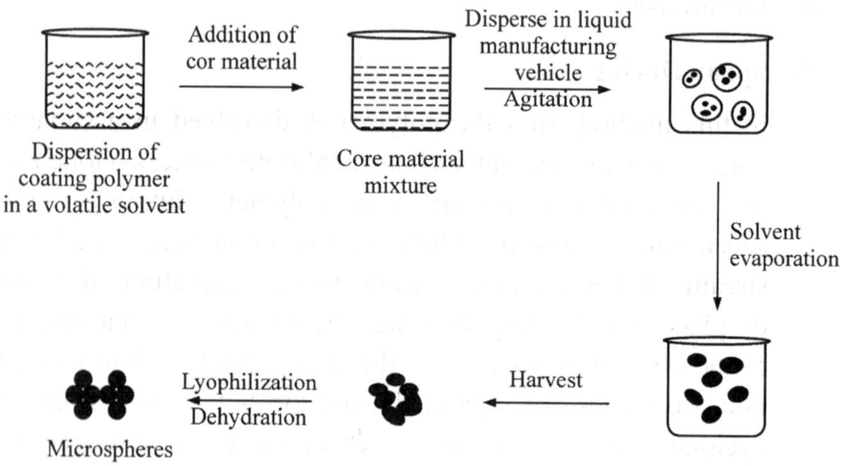

Fig. 5.2 Preparation of Microspheres by Solvent Evaporation.

- This method is mainly used for the preparation of microcapsules.
- In this method, the microcapsule coating is first dispersed in a volatile solvent and is immiscible with the liquid manufacturing volatile phase. The core material to be microencapsulated is dissolved in the coating polymer solution. Then the core material mixture is dispersed in the liquid manufacturing vehicle phase by means of agitation. After that the mixture is heated to evaporate the solvent and to shrink the polymer around the core.
- The matrix type microcapsules are prepared by dissolving the core material in the coating polymer solution.

- Solvent evaporation involves the formation of an emulsion between polymer solution and an immiscible continuous phase.

  E.g., Microspheres of hyaluronic acid, chitosan-glutamate microspheres and a combination of the both are prepared by solvent evaporation.

3. Multiple Emulsion Method

In this method the first step involves the formation of primary emulsion (o/w), then prepares o/w/o multiple emulsions by the addition of primary emulsion to external oily phase and is followed by either addition of cross-linking agent (glutaraldehyde) or evaporation of organic solvent.

E.g., Preparation of chitosan microspheres loaded with hydrophobic drug ketoprofen.

4. Precipitation-chemical Crosslinking

The name itself indicates, the process involves the precipitation of the polymer followed by chemical cross linking. Sodium sulphate is used for precipitation and glutaraldehyde is used as cross-linking agent.

This method involves the addition of aqueous polymer solution such as 3% w/v chitosan and 4% w/v glacial acetic acid into agitating medium and stir it continuously to obtain wet microspheres and are filtered, washed and finally dried at room temperature.

5. Thermal Crosslinking

Various concentrations of chitosan solutions were prepared by maintaining a constant molar ratio between chitosan and citric acid. Citric acid is used as cross-linker. The citric acid solution was added then cooled at 0°C and add corn oil followed by thermal cross-linking at 120°C.

6. Ionotropic Gelation

The chitosan solution in acetic acid was introduced drop wise through a needle into different concentrations of magnetically stirred aqueous solutions of tripolyphosphate or some other ionic solutions. Then the beads were removed by filtration, washed with distilled water and dried.

The counter ions used for ionotropic gelation are of 3 types.

(i) Low molecular weight counter ions e.g., pyrophosphate and tripolyphosphate.

(ii) Hydrophobic counter ions e.g., alginate and carragenan.

(iii) High molecular weight counter ions e.g., octyl sulphate and lauryl sulphate.

7. Complex Coacervation

The microspheres are formed due to interionic interaction between oppositely charged polymeric solutions and KCl and $CaCl_2$ solutions. Then the formed microspheres were hardened in the counter ionic followed by washing and drying. The counter ions used for complex coacervation are sodium alginate, sodium CMC, K-carragenan, sodium polyacrylic acid.

8. Wet Inversion

In this method, the chitosan solution in acetic acid was introduced into an aqueous solution of sodium tripolyphosphate counter ion through a nozzle. The formed microspheres were allowed to stand for 1 hour, washed and add 5% ethylene glycol-diglysidyl ether as cross-linking agent. Then the microspheres were washed and freeze dried to form porous chitosan microspheres. The pore structure of chitosan microspheres is modified by changing the pH of coagulation medium.

Evaluation of Microspheres

1. Particle size analysis
2. Bulk density
3. Tapped density
4. Hausner ratio
5. Determination of drug content
6. SEM analysis
7. *In vitro* drug release studies
8. Data analysis

1. Particle Size Analysis

It is performed by using ordinary microscope. The microscope is first calibrated with a stage micrometer. Then the microspheres were mounted on a slide using glycerin and the average particle size is determined by using eye piece micrometer. Then the average particle size is obtained by multiplying the obtained value with calibration factor. From that the average size frequency distributed curve is drawn by plotting the particle range on x-axis and number of particles on y-axis.

2. Bulk Density

Weigh the sample accurately and filled into 100 ml graduated cylinder and the powder was leveled and unsettled volume V_0 finally. The bulk density was calculated by using the formula i.e.

$$\text{Bulk density} = M/V_0$$

Where,

M = Mass of powder taken

V_0 = Apparent unstirred or initial volume

Units for bulk density is g/cm^3.

3. Tapped Density

The sample under test was filled in 100 ml graduated measuring cylinder and is subjected for mechanical tapping by using tapped density tester at a rate of 300 drops per minute for 500 times and finally note the tapped volume V_t.

The tapped density was calculated by the formula i.e.

$$\text{Tapped density (Ta)} = M/V_t$$

Where,

M = Weight of sample powder taken

V_t = Tapped volume

Units for tapped density

4. Hausner Ratio

It is the ratio of tapped density and bulk density.

$$\text{Hausner ratio} = T_d/B_d$$

Where,

T_d = Tapped density

B_d = Bulk density

5. Determination of Drug Content or Entrapment Efficacy

To determine the drug content or entrapment efficacy first the microspheres were dissolved in 0.1 N HCl for 24 hours. To that add 0.5% v/v Tween 80 to increase the solubility and the amount of drug loaded in the microspheres was determined by spectrophotometer at 284 nm. All the experiments were performed in triplicate and values were tabulated.

$$\text{Loading efficacy (\%)} = w_a/w_t$$

Where,

w_a = Actual weight of drug

w_t = Theoretical drug content

6. Sem Analysis

Scanning electron microscopy helps to observe surface morphology of microspheres. In this method, the microspheres were first vacuum dried, coated with gold, palladium and finally observed under microscope.

Applications

- **Nasal delivery**

  E.g., Methylpyrrolidinone chitosan based spray dried microspheres are used as new carriers in metoclopramide nasal administration.

- **Colonic drug delivery**

  E.g., Colonic drug delivery of enteric coated calcium pectinate microspheres.

- **Protein/peptide stability**

  The key to the success of proteins to be prepared as pharmaceutical products is to have in place an efficient drug delivery system that allows the protein drug to gain access to their target sites at the right time and for the proper duration. Four factors must be considered to fulfill this goal: route of administration, pattern of drug release, method of delivery, and fabrication of formulation. Microspheres help to protect proteins because they are not allowed to react with anything until the polymer is degraded, thus minimizing the contact with solutions that could cause the patients to react.

  Examples:

  1. Albumine (prototype).
  2. Vaccines (HI), diphtheria toxoid (DT), tetanus toxoid (TT), and pertusis toxin (PT) in poly (lactic acid) and PLGA microspheres were prepared by spray drying, all antigens were found to maintain their immunogenicity.
  3. Lysozymes.

- **Drug targeting**

  Drug targeting could be the greatest advantage of microspheres. Most drugs are targeted in the body to give desired results either in specific tissues or organs. A good example of how microsphere technology could be implemented is targeting cancer cells in chemotherapy, as drugs and chemical agents attack cancer cells but have a toxic effect on healthy ones which could very easily cause the cells to die.

(a) Passive targeting

Passive targeting depends on the size of microspheres. The lung's capillaries will let the passage of particles less than seven microns through if one wants the drug to be released into the lungs then the correct size would be around ten microns since they would then be captured in the capillaries.

E.g., Carboplatine microspheres.

(b) Active targeting

Intigrin, lectin, immunoglobulins, lipoproteins, monoclonal antibodies, specific peptides and receptor antagonists were used as ligands conjugated with microspheres as leading molecules for precise targeting.

- **Gene delivery**

  Encapsulation of therapeutic agents such as DNA in microspheres protects the agent from enzymatic degradation, enhances tissue specificity due to localized delivery, eliminates the need for multiple administrations and allows for controlled and sustained delivery.

- **Microspheres in diagnostic materials**

  Gamma emitters such as ^{99}Tc and I^{131} have been incorporated with microspheres for diagnostic purposes. Radio labeling of microspheres is usually achieved either during or after their preparations. Although the former method is still more commonly used in medicine, the later is preferred, especially for shorter lived radioisotopes, because stability and logistical problems are minimized in this way.

CHAPTER 6

MUCOADHESIVE DRUG DELIVERY SYSTEMS

Introduction

This chapter considers both local and systemic delivery of drugs and discusses drug delivery systems that are applied to a variety of sites of the oral cavity including the gingiva, buccal, sublingual, and periodontal pocket.

The mucosa of the mouth is very different from the rest of the gastrointestinal tract and morphologically is more similar to skin. Although the permeability of skin is widely regarded as poor, it is not generally appreciated that the oral mucosa lacks the good permeability demonstrated by the intestine. These differences within the gastrointestinal tract can largely be attributed to the organization of the epithelia, which serve very different functions. A simple, single-layered epithelium lines the stomach, small intestine, and colon, which provides for a minimal transport distance for absorbents.

Concept

Adhesion can be defined as the bond produced by contact between a pressure - sensitive adhesive and a surface. For drug delivery purposes, the term 'Bioadhesion' implies attachment of a drug carrier system to a specific biological surface. The biological surface can be epithelial tissue. If adhesive attachment is to a mucus coat, then the phenomenon is referred to as 'Mucoadhesion'. So, Mucoadhesion is a state in which two components (of which one is biological origin) are held together for extended periods of time by interfacial forces.

In biological systems four types of bioadhesion can be distinguished as:

(a) Adhesion of a normal cell on another normal cell

(b) Adhesion of a cell with a foreign substance,

(c) Adhesion of a normal cell to a pathological cell,

(d) Adhesion of an adhesive to a biological substance

Drug Delivery Systems of Mucoadhesives

Mucoadhesive drug delivery systems include the following:

- Buccal delivery system
- Oral delivery system
- Vaginal delivery system
- Rectal delivery system
- Nasal delivery system
- Ocular delivery system

Need of Mucoadhesive

- For sustained release of drug.
- To target drug delivery & also for localized drug delivery.
- To bypass or minimize hepatic first pass metabolism so that drug degradation can be avoided.
- For Prolonged therapeutic effect.
- For high and rapid drug movement through the absorbing tissue.
- To reduce in fluctuation of steady state plasma level.

An ideal dosage form is one, which attains the desired therapeutic concentration of drug in plasma and maintains constant for entire duration of treatment. This is possible through mucoadhesive drug delivery system.

Theories of Mucoadhesion

Various theories governing the concept of mucoadhesion are:

1. Wetting theory
2. Electronic theory
3. Diffusion theory
4. Fracture theory
5. Adsorption theory

1. Wetting theory

The ability of the adhesive to spread on the mucin to influence the intimate contact between the mucoadhesive and mucin and to increase the mucoadhesive strength. The contact angle should be equal or close to zero. The following diagram shows influence of contact angle between mucoadhesive device and mucus membrane.

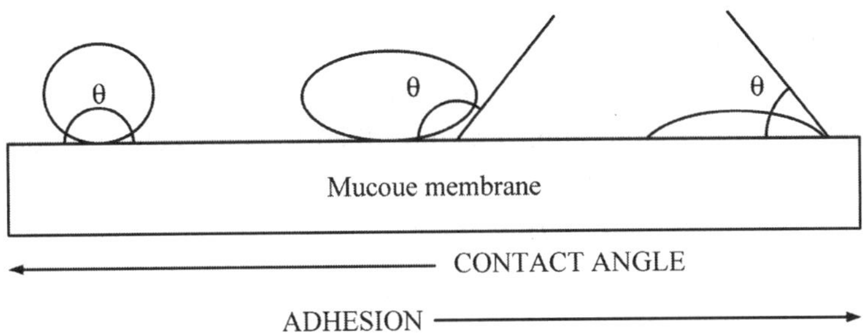

Fig. 6.1 Wetting theory.

2. Electronic theory

The electronic theory says that both mucoadhesive and biological materials possess opposing electrical charges. Thus, when both materials come into contact with one another, they transfer electrons which lead to the development of a double electronic layer at the interface. The attractive forces within this electronic double layer determine the strength of mucoadhesive.

3. Diffusion theory

According to this theory, the polymer chains and the mucous mix to a sufficient depth to create a semi-permanent adhesive bond. The exact depth to which the polymer chains penetrate the mucous depends on the diffusion coefficient and the time of contact. This diffusion coefficient, in turn, depends on the value of molecular weight between crosslinks and decreases significantly as the cross linking density increases. The following diagram shows secondary interactions resulting from interdiffusion of polymer chains of bioadhesive device and mucous.

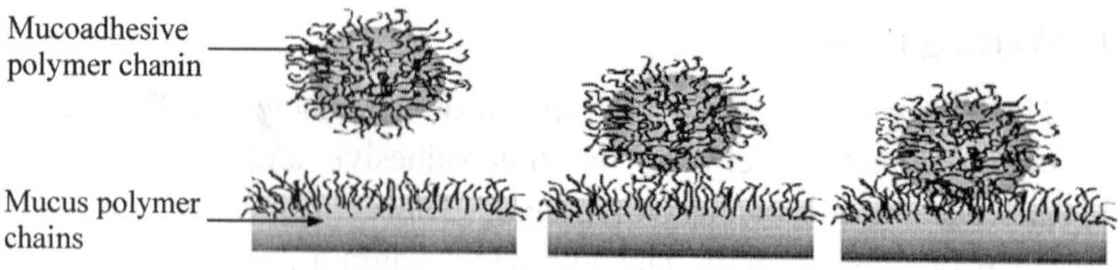

Fig. 6.2 Diffusion theory.

4. Fracture Theory

This theory attempts to relate the difficulty of separation of two surfaces after adhesion. Fracture theory equivalent to adhesive strength is given by:

$$G = (E\,\square/L)^{1/2}$$

Where,

E is the Young's Modulus of elasticity

□ is the Fracture energy and

L is the critical crack length when two surfaces are separated

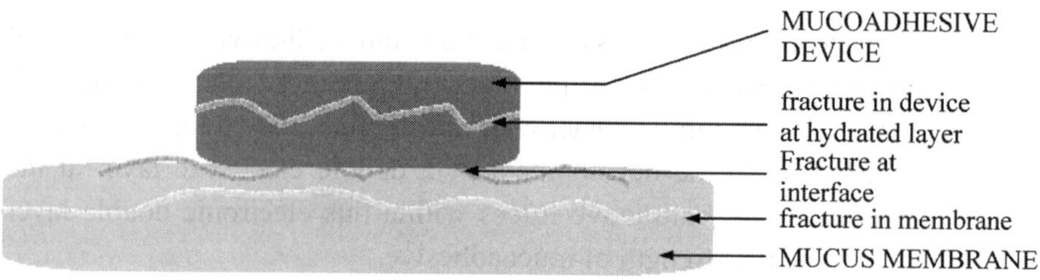

Fig. 6.3 Fracture Theory.

5. Adsorption theory

Attachment of an adhesive to biological tissue on the basis of a group of interactions known as secondary chemical interactions like Van Der Waal's forces, hydrogen bonds, electrostatic attractions or hydrophobic interactions. These forces are considered as the most important in the adhesive interaction phenomenon because though they are individually weak, a great number of interactions can result in an intense adhesion. Two types of chemical bonds resulting from these forces can be distinguished.

(i) Primary chemical bonds of covalent nature, which are undesirable in bioadhesion because their high strength may result in permanent bonds.

(ii) Secondary chemical bonds having different forces of attraction including electrostatic forces, Van Der Waal's forces and hydrogen and hydrophobic bonds.

Steps of Mucoadhesion

1. Contact stage
2. Consolidation stage

These stages are well explained in the following diagram

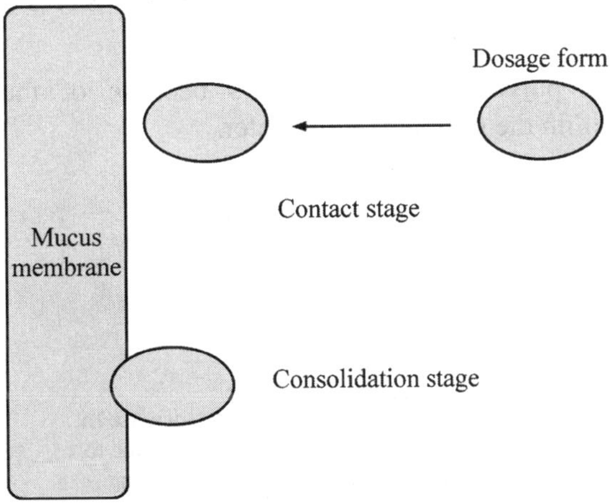

Fig. 6.4 Steps of Mucoadhesion.

Mechanism of Mucoadhesion

Mucoadhesion is mainly based on molecular interactions. The interaction between two molecules is composed of attraction and repulsion. Attractive interactions include Van Der Waal's forces, electrostatic attractions, hydrogen bonding and hydrophobic interactions. Repulsive interactions include electrostatic and steric repulsion. For mucoadhesion to occur, the attractive interaction should be more than non -specific repulsion.

The process of mucoadhesive bond formation has been described in three stages:-

STAGE1: Wetting and swelling of polymer to permit intimate contact with biological tissue.

STAGE2: Interpenetration of bioadhesive polymer chains and entanglement of polymer and mucin chains.

STAGE3: Formation of Chemical bonds between the entangled chains.

Stage 1: Wetting and Swelling of Polymer

Wetting occurs when the polymer spreads over the surface of the mucosal membrane so as to develop an intimate contact with the substrate. Mucoadhesive substrate either adheres to mucous membrane or forms a bond with it with the help of the surface tension and forces existing at the site of adsorption or contact.

Swelling of polymers mainly occurs because of the affinity of components within the polymers with water.

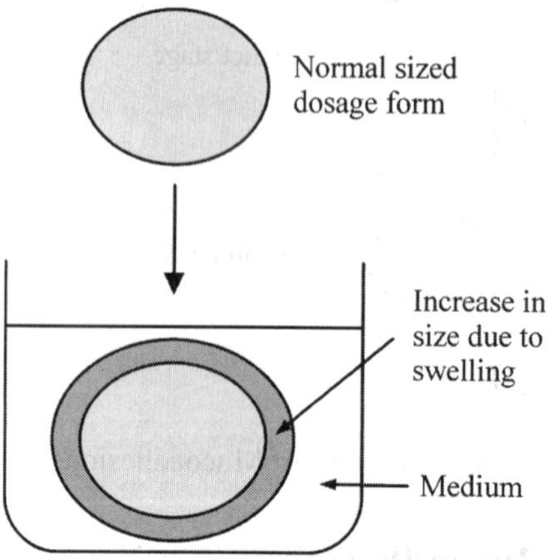

Fig. 6.5 Wetting and Swelling of Polymer.

Stage 2: Interpenetration between the Polymer Chains and the Mucosal Membrane

Interdiffusion and interpenetration which takes place between the mucoadhesive polymer chain and the mucous gel network so as to create a great area of contact. The strength of these bonds depends on the degree of penetration between the two polymer groups. In order to form strong adhesive bonds, one polymer group should be soluble in the other and both polymer types should have similar chemical structure.

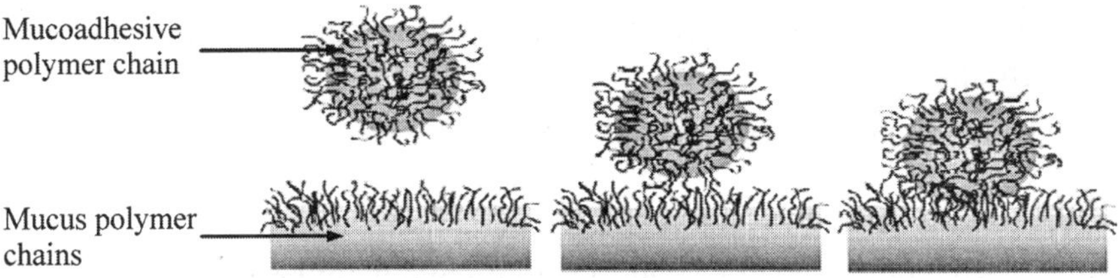

Fig. 6.6 Interpenetration between the polymer chains and the mucosal membrane.

Stage 3: Formation of Chemical Bonds between the Entangled Chains

Entanglement and formation of weak chemical and secondary bonds between the polymer chains and mucin molecule includes primary bonds like covalent bonds and weaker secondary interactions like Van Der Waal's Interactions and hydrogen bonds.

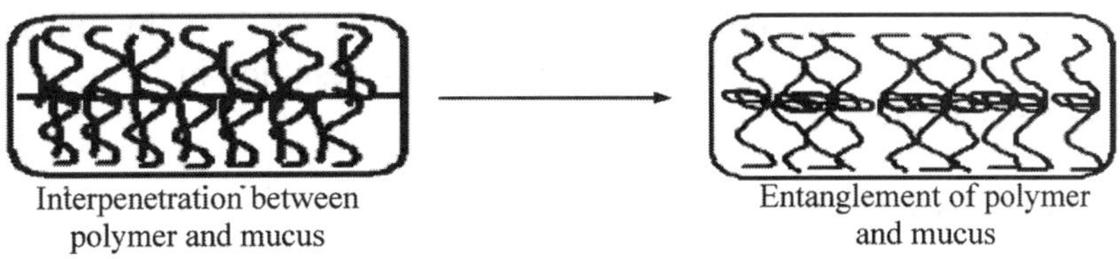

Fig. 6.7 Formation of chemical bonds between the entangled chains.

Mucoadhesive Polymcrs

Various types of polymers are used in mucoadhesives formulation. The polymcrs have the following properties.

Properties

1. It must be loaded substantially by the active compound.
2. It should swell in the aqueous biological environment of the delivery–absorption site.
3. It must interact with mucus or its components for adequate adhesion.
4. When swelled they allow controlled release of the active compound.
5. They are excreted unchanged or biologically degraded to inactive, non-toxic metabolites.

6. It should contain sufficient quantities of hydrogen bonding chemical groups.
7. It should possess chain flexibility.

Polymer Properties

Bioadhesive polymers are categorized into two main properties,

1. Water soluble, linear and random polymers
2. Water insoluble polymers that are swellable networks joined by crosslinking agents.

There are so many properties desirable with mucoadhesive properties of polymers.

Hydrophilic functional groups

The mucoadhesive polymer are generally macromolecules with hydrophilic functional groups such as carboxyl, hydroxyl, amide and sulphate groups which are more favorable in formulating targeted drug delivery system. The polymer interacts with mucus both by physical entanglement and also through chemical bonds, resulting in formation of cross linked network.

Degree of hydration

Hydration is required for the relaxation and interpenetration of polymer chains. Excess of hydration leads to decreased mucoadhesion and/or retention due to the formation of slippery mucilage. Sufficient water is required to properly hydrate the mucoadhesive, expand the gel to create pores of sufficient size for interpenetration.

Chain length

Chain length and its flexibility are important for interpenetration and entanglement with the mucus gel. Increased chain mobility leads to increase inter diffusion and interpenetration of the polymer within the mucus network. Long polymer chains generally lose their ability to diffuse and interpenetrate through mucosal surfaces. Hence it is clearly understood that as the chain length decreases interpenetration increases.

Degree of cross linking

The strength of mucoadhesion decreases with an increase in concentration of the crosslinking agent. Hence an increase in crosslinking density decreases the diffusion coefficient and chain segment flexibility

Polymer concentration

Polymer concentration mainly depends on physical state of the delivery system, with differences between semisolid and solid-state dosage form. For the semisolid state, polymer concentration is low which reduces adhesion and for solid state (dosage forms such as buccal tablets) as the mucoadhesive polymer concentration increases, adhesive strength is increased.

Classification of Mucoadhesive Polymers

I. Based on specifity

1. **The specific bioadhesive polymers**

   These have the ability to adhere to specific chemical structures within the biological molecules. E.g. Lectins, fimbrin.

2. **The nonspecific bioadhesive polymers**

   These have the ability to bind with both the cell surfaces and the mucosal layer. E.g. Polyacrylic acid, cyanoacrylates

II. Based on origin

1. **Synthetic polymers**

   E.g: Cellulose derivatives:-methylcellulose, ethylcellulose, hydroxy-ethylcellulose, poly acrylic acid polymers like polymethylacrylate, polyvinyl pyrrolidone, polyvinyl alcohol.

2. **Natural polymers**

   e.g: Tragacanth, sodium alginate, karaya gum, guar gum, xanthan gum, lectin, soluble starch, gelatin, pectin, chitosan.

Examples of Some Mucoadhesive Formulations

S.no	Drugs	Category	Polymer used
1.	Itroconazole	Antifungal	HPMC
2.	Theophylline	CNS stimulant	Chitosan
3.	Nystatin	Antifungal	Carbapol:HPMC
4.	Lidocaine	Local anesthetic	Antifungal
5.	Prednisolone	Immunosuppressant	Chitosan
6.	Propranolol HCl	Beta blocker	Sodium alginate and carbapol
7.	Insulin	Hypoglycemic agent	Sodium carboxy methyl cellulose

Advantages of Mucoadhesives

- Prolonged residence time is observed at the site of drug action or absorption.
- Localization of drug action of the delivery system can be achieved at a required target site.
- A direct contact with intestinal cells enhances absorption.
- Ease of administration.
- A Best suited delivery system for drugs with first pass metabolism, acid sensitive drugs and those drugs with poor bioavailability.
- This can also be administered to unconscious patients and termination of therapy is easy. {except gastrointestinal}
- A significant reduction in dose can be achieved there by reducing dose related side effects.
- It offers a passive system of drug absorption and does not require any activation.
- Systemic absorption is rapid.
- This route provides an alternative basis for the administration of various hormones, narcotic analgesic, steroids, enzymes, cardiovascular agents etc.
- Contributes improved patient convenience and compliance due to less dosing frequency, shorter treatment period.
- Safety margin of high potency drugs is increased due to better control of plasma levels.
- Maximum utilization of drug is seen enabling reduction in total amount of drug administered.

Disadvantages

- Leads to the occurrence of local ulcerous effects due to prolonged contact of the drug possessing ulcerogenic property.
- Lack of a good model for *in vitro* screening to identify drugs suitable for such administration.
- Patient acceptability in terms to taste, irritancy and mouth feel is to be checked.

Limitations

- Drugs that cause irritation to the oral mucosa and those with a bitter or unpleasant taste, odour, cannot be administered by this route.
- Drugs, that are unstable at buccal pH, cannot be administered by this route.
- Only those drugs, which are absorbed by passive diffusion and those with small dose requirements, can be administered by this route.
- Eating and drinking may become restricted when buccal mucoadhesive is attached in oral cavity.
- Swallowing of the formulation by the patient is sometimes possible.
- Over hydration of polymer leads to the formation of slippery surface and structural integrity of the formulation may get disrupted by the swelling and hydration of the bioadhesive polymers.

Evaluation of Mucoadhesives

In vitro and *ex vivo* Tests

There are many methods of *in vitro* and *in vivo* tests for characterizing mucoadhesive properties such as:

- Methods determining tensile strength
- Methods determining shear stress
- Thumb tack method
- Swelling properties
- *In vitro* drug release studies
- Mucoretentability studies

In vitro/ex vivo tests are very important in the development of a controlled release bioadhesive system because they help in the studies of permeation, release, compatibility, mechanical and physical stability, superficial interaction between formulation and mucous membrane and strength of the mucoadhesive bond. These tests can simulate a number of administration routes including oral, buccal, periodontal, nasal, gastrointestinal, vaginal and rectal. The *in vitro* and *ex vivo* tests which are most prevalent in the literature are described below.

Techniques utilizing gut sac of rats

The everted gut sac technique is one of the *ex vivo* method. This is used since 1954 to study intestinal transport. This method is easy to reproduce and perform in almost all laboratories. A small segment is removed from the rat's intestinal tissue. One of its ends is sutured and filled with saline. The sacs are then introduced into tubes containing the system under analysis at known concentrations. It is stirred, incubated and then removed. The percent adhesion rate of the release system onto the sac is determined by subtracting the residual mass from the initial mass.

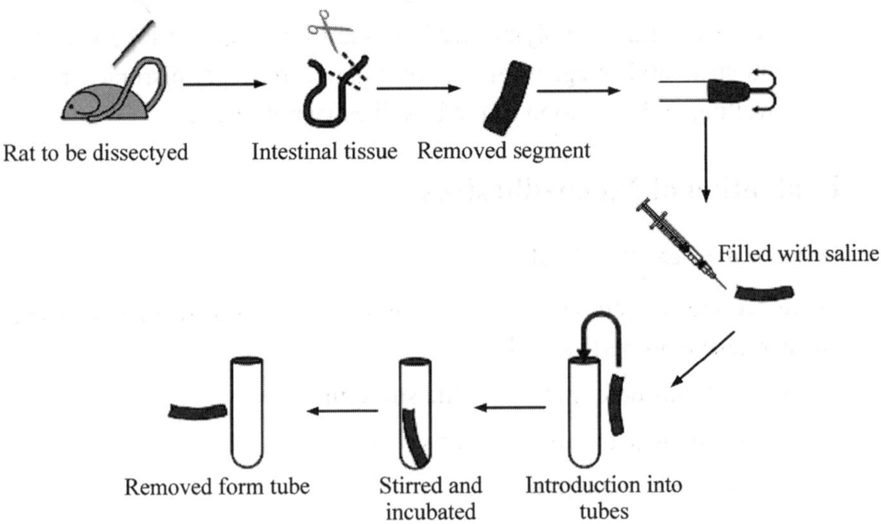

Fig. 6.8 Everted gut sac procedure.

Tests measuring mucoadhesive strength

Mucoadhesive strength is the force that is required to break the binding between the model membrane and the mucoadhesive attached to it. Depending on the direction in which the mucoadhesive is separated from the substrate, we obtain the detachment, shear, and rupture tensile strengths as indicated in Figure.

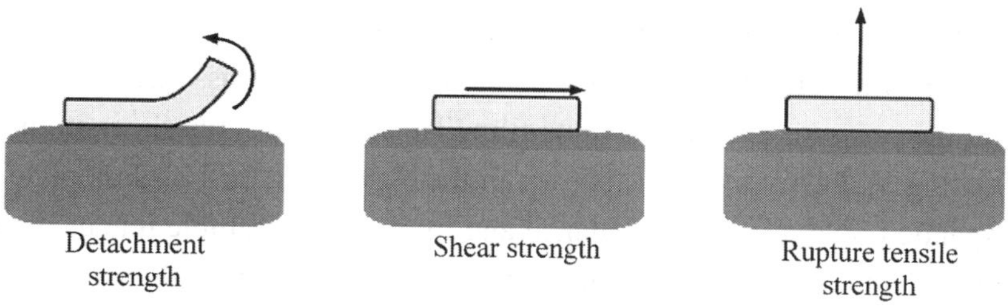

Fig. 6.9 Tests measuring mucoadhesive strength.

The equipment used for this is a texture analyzer (shown in the figure) or a universal testing machine. In this test, mucoadhesive strength is measured, by using a disc composed of mucin, a piece of animal mucous membrane or generally porcine nasal mucus or intestinal mucus from rats. From the results, a plot can be drawn between the force and distance. From the plot we can get mucoadhesive strength, tensile work (area under the curve during the detachment process), peak force and deformation to failure.

In addition to rupture tensile strength, the texture analyzer (shown in the figure) also analyzes the texture of the formulation and other mechanical properties of the system.

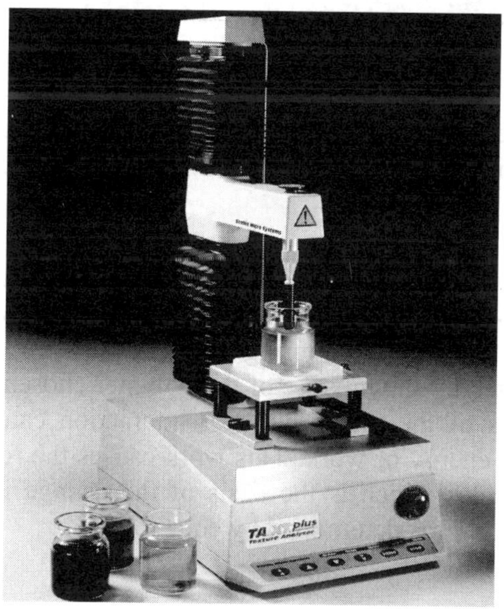

Fig. 6.10 Texture analyzer.

Mucoadhesion strength can also be measured in terms of shear strength (the force required to separate two parallel glass slides covered with the polymer and with mucus film. This test measures shear strength by using Wilhemy's model (shown in the figure). In this, a glass plate is suspended by a microforce balance and immersed in sample mucus under controlled temperature. The shear strength of the sample is then measured under constant experimental conditions.

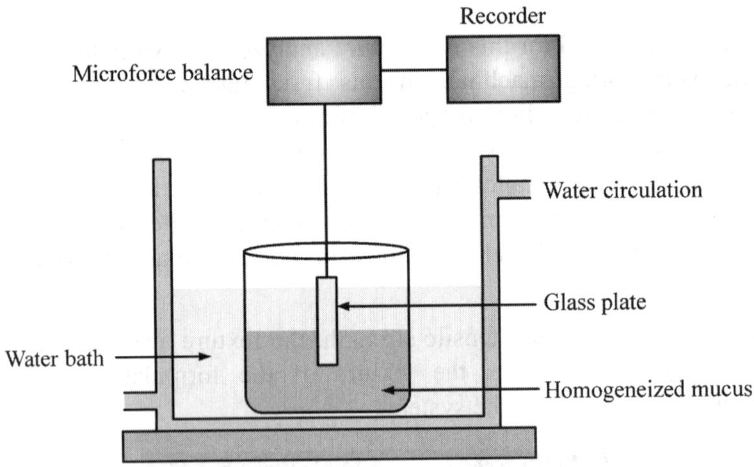

Fig. 6.11 Wilhemy's model.

Wilhemy's plate technique, or the microforce balance technique, can also be modified to measure the specific adhesion force of microparticles.

Rheological methods

These methods are carried out *in vitro* by viscosimetric assays to analyze the formulation-mucin interaction. From this test, one can estimate the mucoadhesion force by monitoring the viscosimetric changes of the system constituted by the mixture of the polymer chosen and mucin. The energy of bonds of the mucin-polymer interaction can be transformed into mechanical energy or work. This work causes the rearrangements of the macromolecules which is the basis of the change in viscosity. The bioadhesion force, F, is determined by equation:

$$F = \eta b \sigma$$

Where, σ is the shear gradient

ηb is bioadhesion component

Tests analyzing molecular interactions involved in mucoadhesion

The general problem arising from methods that show the adhesion force and from the rheological methods is that the mucoadhesive response is seen macroscopically while the interactions occur at a microscopic level.

So by using low frequency dielectric spectroscopy, one can attempt to study gel-mucus interactions near the molecular level. It evaluates the physicochemical interactions between molecules and glycoproteins of the mucus at the interface. This technique provides information about the compatibility between mucus and mucoadhesive system by means of the evaluation of the movement of the charged particles. The dielectric measures reveal information about the gel and the mucous membrane separately, and about the interface between them.

Since the mucoadhesion process is based on the interactions between the mucus layer and the mucoadhesive polymer, it is highly depends on the molecular structure, including its charge. It is well known that glycoproteins molecules, which form the mucus structure, are negatively charged at physiological pH. By zeta potential, the polymer-mucin electrostatic interactions can be understood. In this test, the mucin particles are suspended in an appropriate buffer and mixed with a solution of the polymer. If the polymer added changes the zeta potential value of the mucin particles, it suggests the existence of greater affinity between polymer and mucin particles.

Another technique which can be applicable to evaluate molecular interactions is the optical biosensor, or resonant mirror biosensor technique.

In vivo Method of Evaluation

These are important to asses the true mucoadhesive potential in case of oral mucoadhesive drug delivery.

- Use of radioisotopes
- Use of gamma scintigraphy
- Use of pharmacoscintigraphy
- Use of electron paramagnetic resonance(EPR) oximetry
- X ray studies
- Isolated loop technique

The GI transit time can be measured by using one of the many radio opaque markers like barium sulphate which is coated to the bioadhesive dosage form so as to assess the GI transit by means of x-ray inspection. By means of gamma scintigraphy both the distribution and retention can be studied. *In vivo* mucoadhesion was determined by ascertaining there residence time of the fluorescently tagged thiomer on intestinal mucosa of rats after 3 hours.

CHAPTER 7

ORAL CONTROLLED RELEASE DRUG DELIVERY SYSTEMS

Introduction

Controlled drug delivery is drug delivery that delivers the drug at a predetermined rate, for locally or systemically, for a specified period of time. In this drug delivery, continuous oral delivery of occurs at predictable and reproducible kinetics for predetermined period throughout the course of GIT. Drug is being delivered at therapeutically effective rate to desirable site in this type of drug delivery.

Advantages

- Oral route is more commonly used one and it is more convenient form of drug delivery systems.
- The designing of oral dosage forms is very easy.
- It is much deviated from sterility problems.
- I t plays an important role in delivery of drugs for prolonged time.
- It improves patient compliance by reducing frequency of administration.
- Reduced side effects by reducing dose amount (low dose).
- **Therapeutic advantage:** Avoid fluctuations of plasma drug level with OCDDS due to slow and continuous release of drug from its dosage form and it maintains constant drug level.
- It provides better stability of drug.
- It reduces the health care cost.

Disadvantages

- Dumping of the dose.
- Self medication is not possible.
- Reduced potential for accurate dosing is observed.
- Stability of the formulation is a problem.

Basic Mechanisms Involved in OCDDS

The mechanisms mainly involved in this to achieve the slow and controlled release rate of drug in GIT are diffusion, dissolution, combination of both diffusion and dissolution and osmotic pressure controlled release systems.

Classification of OCRDDS

Based on the mechanism OCRDDS, they are classified in to 3 types:

1. Dissolution controlled release
    - Dissolution control in encapsulation
    - Dissolution control in matrix
2. Diffusion controlled
    - Reservoir devices
    - Matrix devices
3. Combination of both

Dissolution definition

Dissolution is defined as the solubilisation of a substance in given solvent. It also involves mass transfer from solid to liquid. Here Rate determining step is diffusion from solid to liquid.

Theories to Explain Dissolution

1. Diffusion layer theory
2. Surface renewal theory
3. Limited solvation theory

Dissolution is given by Noyes-Whitney equation:

$$\mathbf{dc/dt = kD.A\ (Cs - C)}$$

$$\mathbf{dc/dt = D/h\ A.\ (Cs - C)}$$

dc/dt = Dissolution rate.

k = Dissolution rate constant (1st order).

D = Diffusion coefficient/diffusivity

Cs = Saturation/ maximum drug solubility.

C = Con. of drug in bulk solution.

Cs-C = concentration gradient.

h = Thickness of diffusion layer.

1. Matrix Type

- It is also referred as Monolith dissolution controlled system.
- Here dissolution phenomenon is mainly controlled by following ways:
  1. Alteration of porosity of tablet.
  2. Decrease in wettability.
  3. Dissolution at slower rate.
- It follows First order drug release.
- Drug release is determined by dissolution rate of polymer.
- E.g., Dimetan extencaps, Dimetapp extentabs.

2. Encapsulation

- It is also referred as coating dissolution controlled system.

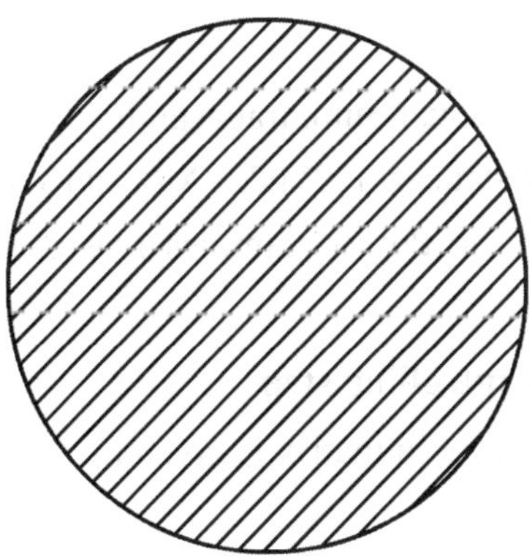

Fig. 7.1 coating dissolution controlled system.

- Dissolution rate of the coating depends upon stability and thickness of coating.

- Coating masks various propertied like color, odor, taste, minimising GI irritation.
- One of the microencapsulation methods is used in this method.
- E.g., Ornadespansules, Chlortrimeton Repetabs

Diffusion

Definition

Drug molecules diffuse from a region of higher concentration to lower concentration until equilibrium is attained. And it was directly proportional to the concentration gradient across the membrane. It is major process for absorption and it doesn't require any energy.

1. Matrix Diffusion Types

Rigid Matrix Diffusion

Materials used in this type are insoluble plastics such as PVP and fatty acids.

Swellable Matrix Diffusion

1. Also referred as glassy hydrogels. These are popular for sustaining the release of highly water soluble drugs.
2. Materials used here are hydrophilic gums.

For example,

- Natural substances: Guar gum, tragacanth.
- Semisynthetic substances: HPMC, CMC and xanthum gum.
- Synthetic substances: Polyacrilamides like glucotrol XL, procardia XL.

Matrix system Rate controlling step

Diffusion of dissolved drug in matrix is given by following equation

Higuchi equation

$$Q = DE/T\ (2A.E\ Cs)Cs.t)^{1/2}$$

Where,

Q is amount of the drug released per unit surface area at time t.

D is diffusion coefficient of drug in the release medium.

E is porosity of matrix.

Cs is solubility of drug in release medium.

T is tortuosity of matrix.

A is concentration of the drug present in the matrix per unit volume.

2. Reservoir System

- Also called laminated matrix device.
- Consists of a hollow system which contains an inner core surrounded in water insoluble membrane.
- Polymer can be applied either by coating or microencapsulation.
- Mechanism of rate controlling in this is partitioning into membrane with subsequent release into surrounding fluid by diffusion.
- Commonly used polymers are HPMC, ethyl cellulose and polyvinyl acetate. E.g., Nico-40, Nitro-bid.

Rate controlling steps

- Polymeric content in coating,
- thickness of coating,
- hardness of microcapsule.

Dissolution and Diffusion Controlled Release system

- Drug is encased in a partially soluble membrane.
- Pores are created due to dissolution of parts of membrane. Pores permits the entry of aqueous medium into core.

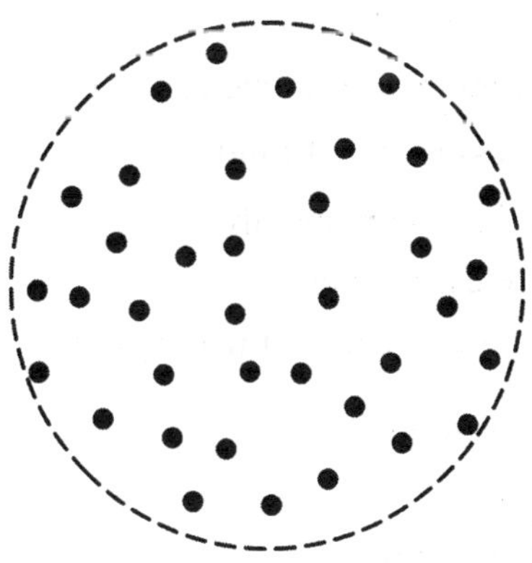

Fig. 7.2 Dissolution and Diffusion Controlled Release system.

- Diffusion of dissolved drug out of system.
- E.g., Ethyl cellulose and PVP mixture dissolves in water and form insoluble water membrane.

Osmotic Pressure Controlled Release Systems

Definition

Movement of solvent molecules from a region of lower to higher concentration. The passage of solvent into a solution occurs through semipermeable membrane.

Osmotic Pressure Controlled System

- It shows zero order release.
- Drug may be either osmotically active or combined with an osmotically active salt (E.g., NaCl).
- Semipermeable membrane is usually made from cellulose acetate.
- This is more suitable for hydrophilic drug.
- E.g., Glucotrol XL, Procardia XL,

Equation

$$\mathbf{(Q/t)\ z = Pw\ Am/\ hm\ (\pi s - \pi e\ )}$$

(Q/t) is Rate of zero order drug release.

Pw is water permeability,

Am is effective surface area

hm is thickness of semipermeable membrane.

πs is osmotic pressure of saturated solution of drug or salt in system

πe is osmotic pressure of GI fluid.

Modifications made in this are mainly

1. Immediate release system.
2. Osmotically active compartment system

Immediate release systems

Steps involved in this are

1. Activation of the sytem

2. Dividing the required dose into two compartments of which one third is immediate release and two third is controlled release. Semi permeable membrane is involved.

   E.g., Phenyl propanolamine.

Osmotically active system

1. Here two compartments involved in a sytem are separated by movable partition.
2. The osmotically active compartment absorbs water from GIT which leads to the creation of osmotic pressure.
3. Due to osmotic pressure the Partition moves upward and then drug is released.

   E.g., Nifedipine.

Different Types of Formulatons

1. Ion-Exchange Resins

Resins are water-insoluble materials containing anionic or cationic groups in repeating positions on the resin chain.

1. The drug-charged resin is prepared by mixing the resin with drug solution either by repeated exposure of the resin to the drug in a chromatographic column or by keeping the resin in contact with the drug solution for extended periods of time. Polymer-coated drug-resin design Polymer-coated drug-resin dispersion.
2. The drug-resin is then washed to remove contaminant ions and dried to form particles or beads. When a high concentration of an appropriately charged ion is in contact with the ion-exchange group, the drug molecules is exchanged and diffuses out of the resin to the bulk solution.
3. pH-Independent formulations:

   The granules are designed for the oral controlled release of basic or acidic drugs at a rate that is independent of the PH in the GI tract. They are prepared by mixing a basic or acidic drug with one or more buffering agents, granulating with appropriate pharmaceutical excipients, and finally, coating with a gastrointestinal fluid permeable film-forming polymer. When the GI fluid permeates through the membrane, the buffering agents adjust the fluid inside to a suitable constant pH, thereby rendering a constant rate of drug release.

2. Altered Density Formulations

- Several approaches have been developed to prolong the residence time of drug delivery systems in the GI tract. One such approach is the bioadhesion approach, which is based on the adherence of bioadhesive polymers to the mucin/epithelial surface of the GI tract.
- The other approach is to alter the formulation's density by using either high or low density pellets.

High-density approach

- In this approach, the density of the pellets must exceed that of normal stomach content and should therefore be at least 1.4. In preparing these formulations, drug can be coated on a heavy core or mixed with heavy inert materials such as barium sulphate, titanium dioxide, iron powder and zinc oxide.
- The weighed pellet can then be covered with a diffusion controlled membrane.

Low-density approach

- Globular shells which have an apparent density lower than that of gastric fluid can be used as carrier of drug for sustained release purposes. Polystyrol, poprice, and even popcorn are as carriers.
- The surface of these empty shells is undercoated with sugar or with a polymeric material such as methacrylic polymer and cellulose acetate phthalate. Drug delivery system with flotation chamber.
- The undercoated shell is then coated by a mixture of drug with polymers such as ethyl cellulose and hydroxy propyl cellulose. The final product floats on the gastric fluid for a prolonged period, while slowly releasing drug.

Types of Tablets

1. Matrix Tablets

There are three types of matrix tablets:

1. Hydrophilic matrices
2. Plastic matrices
3. Fatwax matrices

2. Ion-Exchange Resin Tablets

1. The tablets are commonly developed by embedding the drug molecules in the ion-exchange resin matrix and then this core is coated with a semi permeable coating material such as ethyl cellulose.
2. The most widely used and safe ion exchange resin is divinyl benzene sulphonate. Drugs that were prepared using cationic exchange resins include codeine, ephedrine.

3. Film Coated Tablets

1. Simple coating of the tablets by the polymers can control the dissolution rate of drug from the tablets.
2. Various polymers such as H.P.M.C, E.C and Eudragit polymers have been used for the polymeric film coating of the tablets. These polymer coatings act as diffusion controlling membranes.
3. The selection of coating material will decide the dissolution rate of the drug molecules. Plasticizers also accelerate the drug release.

4. Floating Tablets

1. Floating tablets can be used for local action in the proximal GIT. Poorly soluble and unstable as well as poorly absorbable drugs are suitable for floating dosage units.
2. These systems are retained in the stomach for prolonged time due to their floating property.
3. Floating dosage forms have been developed to float over GI fluids and to release the drug over a desired period of time.
4. Increased GI time is the consequent property of floating tablets.
5. Drugs suitable to be formulated as floating tablets morphine, captopril and diazepam.

5. Enteric Coated and Delayed Release Tablets

1. The materials used for enteric coating are acid impermeable polymers. An ideal enteric coating should dissolve at a pH slightly lower than 7.
2. Various solvents such as alcohol, acetone, and methylene chloride have been used to dissolve the enteric coating polymers.

For example,

1. Aspirin given as enteric coated tablets prevents gastric irritation.
2. Erythromycin degrades in gastric pH. Hence, it is given as prodrug/enteric coated formulation.

The enteric coating formulation needs higher pH 6.5-7.0 for disintegration of coating, which delays the release of erythromycin beyond the absorption site. The efficacy of formulation depends on selection of suitable polymer.

6. Osmatic Tablets

1. The osmotic tablets are comprised of a drug contained in a rigid, semi permeable membrane in which an aperture is created by a mechanical drill or laser beamcontroled release of an tablet can beachieved by constructing a tablet with two basic components
2. These systems are suitable for the soluble drugs such as indomethacin and metoprolol.
3. Semi permeable polymers such as cellulose acetate and polyurethane are commonly used for osmotic tablet formulations.

7. Mucoadhesive Tablets

1. They are developed for administration of drug to the oral cavity. Buccal and sublingual mucosa offer for mucoadhesion.
2. Drug can be applied locally and removed. Synthetic polymers including cellulose derivatives and polyacrylic acid are used for preparation of bioadhesive tablets.

8. Swellable Tablets

1. The dosage forms can be developed in a size, which can be swallowed and when they reach the stomach fluid swell quickly and attain considerably large size.
2. In addition to swelling, gel forming property of the polymer can retain the drug molecules within the dosage form thereby sustaining the release of drug from the formulation.

9. Multiple Unit Tablets

1. They contain subunits that are homogenous or heterogenous.
2. They are prepared by combining the immediate release portion and controlled release portion.

10. Microcapsules and Microspheres

1. The mechanism of drug release is either dissolution or diffusion of the drug and the formulation are either microencapsulated or matrix. Various methods including interfacial polymerization, coacervation/phase separation are used to prepare microparticles.
2. The coating material should be selected depending on the material to be coated and the release characteristics desired.

Evaluation

1. Physical Evaluation

Color	:	Any imperpection.
Texture and membrane size	:	Size height and radius
Scanning electron microscope	:	Drug content
Dissolution behavior	:	Stability studies

2. Weight Variation and Hardness Determination

Weight variation was determined by weighing 20 tablets of each formulation on an electronic balance. The hardness of 10 tablets was measured using hardness tester prior to coating.

3. Scanning Electron Microscopy

Coating membranes of formulation obtained before and after complete dissolution of core contents were examined for their porous morphology by scanning electron microscope membrane were dried at 45°C for 12 hours.

4. Effect of PH

To study the effect of pH and to assure a reliable performance of the developed formulation independent of pH, *in vitro* release studies were conducted in media of different pH. The release media was SGF (pH1.2), acetate buffer (pH 4.5), and simulated intestinal fluid (6.8).samples were analyzed by HPLC and stored between sheets of wax paper in desiccators until examination.

5. Biopharmaceutical Considerations

The *in vitro* test was conducted for the purposes of

1. To provide the process control and stability determinations of the relevant release characteristics.
2. To facilitating certain regulatory determinations and judgments, concerning minor formulation changes, site of manufacturing change, etc.
3. The key elements for the dissolution are:
   (a) Reproducibility of the method.
   (b) Selection of media.
   (c) Maintainence of sink conditions.
   (d) Dissolution rate as a function of pH ranging from pH 1 to 8. Selection of the most discriminating variables as the basis for the dissolution test and specification.

6. Dissolution

The dissolution procedure should establish for

1. Lack of dose dumping – indicated by a narrow limit on the one hour dissolution specification.
2. Controlled release characteristics – by employing additional sampling windows overtime.
3. Complete drug release – indicated by 75-80% minimum release specifications at the last sampling interval.
4. Dosage form pH dependence/independence indicated by percent dissolution in water, appropriate buffer, gastric and simulated intestinal fluid.

In vivo Bioavailability Data

A. General

B. Specific types of *in vivo* studies are:

   1. Fasted single-dose studies:
   2. Post prandial study:
   3. Multiple-dose study - state studies:

CHAPTER 8

PRODRUGS

Introduction

The use of prodrug limit the unfavorable characteristics of active drug constituent which prevent them from permeating biological barriers such as intestinal mucosa, lack of stability against enzymatic degradations of API. Prodrug strategy is used as a biochemical approach to face various barriers which can hinder drug delivery including solubility to low solubility of poorly soluble drugs.

Classification of Prodrugs

Prodrugs are classified into two broad categories. They are

1. Carrier linked prodrugs
2. Bioprecursors

Carrier Linked Prodrugs

Carrier linked prodrugs generally consists of the attachment of a carrier group to the active drug. This is so done to modify its physicochemical properties and then subsequent enzymatic or non-enzymatic mechanism to release the active drug moiety.

Types of carrier linked prodrugs

Based on the nature of the carrier used

1. Double prodrug
2. Macromalecular prodrug
3. Site specific prodrug
4. Mutual prodrug

1. Double prodrug is also referred as pro-prodrugs or cascade-latentatiated prodrugs. In this type, prodrug is further derivatized. It is derivatized in such a fashion that only enzymatic conversion to prodrug is possible before the latter can cleave to release the active drug.
2. Macromolecular prodrug is a type where macromolecules such as polysaccharides, cyclodextrins, proteins, peptides and polymers are used as the carriers.
3. Site specific prodrug is a type where a carrier acts as transporter of active drug to the specific targeted site.
4. Mutual prodrug is a type of prodrug where the carrier itself is another biologically active drug instead of some inert molecules. Therefore, a mutual prodrug is consisting of two pharmacologically active agents which are coupled together so that each acts as pro-moiety for the other agents and vice-versa. The carrier that is selected for this purpose may have the same biological action as that of the parent drug and thus ensuring some additional benefits .

The carrier may be useful in the following ways:

1. It may help to target the parent drug to a specific site or organ or cells.
2. It may improve site specificity of a drug.
3. It may overcome side effects of the parent drug.

Ideal characteristics for carrier linked prodrug

1. The linkage between the drug and carrier should usually be a covalent bond. As a rule, the prodrug itself must be inactive or less active than parent drug.
2. The linkage should be bioreversible.
3. Prodrug and the carrier released after *in vivo* enzymatic or nonenzymatic attack should be nontoxic.
4. Generation of active form must takes place with most rapid kinetics to ensure effective drug levels at the site of action.
5. Bioavailability of carrier linked prodrug is modulated by using a transient moiety.
6. Bioactivation process is exclusively hydrolytic and sometimes a redox system.
7. Ideal carrier should be a without intrinsic toxicity.

8. It should be non-immunogenic and non-antigenic, should possess a suitable number of functional groups for drug attachment and adequate loading capacity.
9. It should be stable to chemical manipulation and autoclaving.
10. It should be easy to characterize and should mask the liganded drugs activity until release of active agent at the desired site of action
11. In mutual prodrug approach, the carrier should have some biological activity of its own.

Pharmacosomes

Prodrugs becomes amphiphilic, if the hydrophilic drug is bonded to lipophilic promoiety drug is conjugated to the lipophilic promoiety by covalent bond. If the lipophilic part is a long chain mono- or di-glyceride, then the prodrugs acquires mesomorphic lyotropism and arrange into hexagonal phases. The word mesomorphism means that the system may exists in several structurally different phases . If it occurs by lipid water interaction these are called lyotropic. Prodrugs can be prepared by esterification of the drug free carboxyl group and hydroxyl group of the lipid.

HYDROPHILIC PART OF LIPID + LIPOPHILIC PART

↓

AMPHIPHILIC PRODRUG

The word pharmacosomes was coined by Vaizoglu et al. Pharmacosomes are defined as the colloidal dispersion of the prodrug which can exists in different forms like ultrafine vesicular, micellar or hexagonal aggregates which depends on the chemical structure of drug-lipid complex like niosomes. It is also alternative to liposomes. But unlike liposomes, pharmacosomes do not show leakage of drug storage because the release of drug occurs by hydrolysis.

The formulation of pharmacosomes prodrugs in the form of vesicle decreased many problems like leakage. It avoids the steps of removing the free drug from entrapped drug, where as in the liposome gel filtration, dialysis, ultracentrifugation are techniques used to separate free drug.

Preperation of Pharmacosomes

These are prepared by self-dispersion through moderate mixing and sonication. The preparation methods include;

1. Hand shaking method
2. Ether injection method

In the hand shaking method, the drug is made to form complex with lipid carrier, this complex is made into a film which is dried and deposited in a round bottom flask on addition of aqueous medium vesicular suspension is formed.

In ether injection method, vesicles are formed when organic solution of drug-lipid complex injected into the hot aqueous medium.

Stability of Pharmacosomes

They does not show leakage of drug during storage even no hydrolysis even after 2 months of storage. E.g., Pindolol glycerol monosterate, but the hydroxyl group present in pharmacosomes are very sensitive to electrolytes, on interaction aggregates are found. Newly developed dry granular propharmacosomes show long storage life.

Applications of Pharmacosomes

The cell membranes of our body consists of phosphatidylcholine, phosphatidyl-ethanolamine, ceramides, sphingomyelin which are similar to lipid prodrugs. These pharmacosomes can interact with biomembranes and transfer the active ingredient. The interaction of pharmacosomes with biomembrane can change the phase transition temperature of biomembranes there by enhance the permeation by increasing the membrane fluidity.

Taxol is synthesized by covalent binding of drug with the fatty acid chain of a phospholipids , glyceride , ceramide. Any linkage used for binding of drug with fatty acid should cleave in the body. Taxol was synthesized by mixing 1-oleoyl-2-{n-4-{4'-0=(9' taxyl)-succinyl}-11- aminoundecanoyl}-L-α-phosphatidylcholine and the mixture was dissolved in 10% methanol in benzene and then lyophilized. Then the resultant powder is dispersed in a buffered saline and particle size is adjusted to 90 nm.

Dermatan sulfate showed tensioactivity. The vesicles formed in nanometers shows improved oral bioavailability. Pindolol is glycerol monosterate ester of a weak base. This prodrug shows structural similarity with lysolecithin. It is prepared by film method and ether injection method . Particle size of pharmacosomes is determined by transmission electron microscopy (TEM) and photon correlation of unchanged pindolol than free pindolol administration.

Bupranolol is prepared by covalent binding of bupranolol with 1,3-dipalmitotyl-2-succinyl-glycerol. The prodrug formed resemble the phospholipids showed continuous swelling with increasing water content and in excess water , thermodynamically vesicles are formed along with oligomeric vesicles.

Antibody Based Targeting

Monoclonal antibodies are the better way of targeting tumour cells in which prodrug is activated at tumour site. They are antibody-drug conjugate and antibody directed enzyme prodrug therapy (ADEPT). Antibody-drug conjugate are macromolecular prodrugs synthesized by covalent binding of cytotoxic agent to monoclonal antibodies which react with tumour antigens.

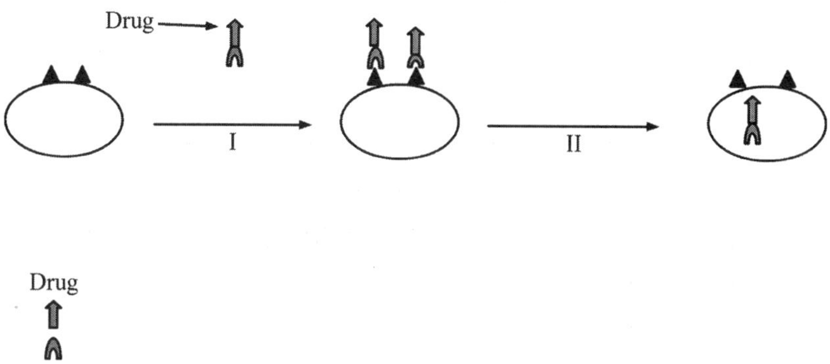

Fig. 8.1 I-antibody drug conjugate preparation
II- internalization and cleavage of antibody drug complex.

Immuno-conjugate binds to specific cells leading to cleavage and drug from antibody intracellularly . Example - metotrexate, BR-6-Doxorubicin.

Disadvantages

1. Inefficient internalization of antigen antibody complex
2. Insufficient cleavage of the linker moiety to release free drug
3. Heterogenecity in the expression of antigen by cancer cells.

In ADEPT, enzyme is covalently linked to a MAb, which selectively binds to its corresponding tumour associated antigen, the antibody enzyme conjugate is localized in the tumour and cleared from nontargeted sites, a prodrug which is a substrate for the enzyme is administered . The targeted enzyme converts the prodrug into active drug at the tumour site. This approach is still under clinical trials for proof.

The third approach is administration of Ab-E first, then a galactosylated anti Ab-E antibody is administered to clear nontumour associated circulating Ab-E and then prodrug is injected.

Rationale for Prodrug Formulation

The process of any drug development occurs in 3 phases which are pharmaceutical, pharmacokinetic, pharmacodynamic. Prodrug concept is used in pharmacodynamic phase where the drug receptor interaction is involved.

In the pharmaceutical phase, following properties of drug has to be considered.

1. Physical properties of the drug substances like taste, odor, pain on injection.
2. Physicochemical properties of the drug substance-

Hydrophilicity, lipophilicity, poor stability.

In pharmacokinetic phase includes ADME process of a drug. The principal barriers are,

1. Incomplete absorption of drug through the biological membranes.
2. Too rapid or too slow absorption or excretion.
3. Low systemic bioavailability due to first pass metabolism.
4. Poor site specificity of the drug.

Some drugs show poor water solubility can result in low bioavailability after oral administration, high water solubility decreases the bioavailability. These problems can be solved by prodrug in which drug substance is covalently liked to an inert chemical entity maintaining its original actions. Enzyme reversible prodrug linkages, includes aliphatic ester, carbonate ester, hemi-ester, sulphate esters, phosphate ester, aminoacids, azolinkages, phosphoramides, amides, carbamates, glucosuronates, β-glucosides, N-acetylglucosamides.

Polar prodrug

↓

Nonpolar drug

↓

Intestinal mucosa

↓

Blood

4. Non polar drug is formed from polar prodrug by colonic bacteria.

Site activated prodrugs for neoplastic cells

As proliferation rate is high in neoplastic cells, enzymatic activity is also high. E.g., [2'deoxy-5-fluorouridine] doxifluoridine is a prodrug of 5-flurouracil.

[2'deoxy-5-fluorouridine]doxifluoridine

↓

5-flurouracil

Site activated prodrug targeting the kidney

Kidney are highly active in metabolism and uptake of gama-glutamyl derivatives of aminoacids, peptides due to the enzymes high gama-glutamyl transpeptidase.

Glucodopa (glutamyl derivatives of L-DOPA)

↓ I

L-DOPA

↓ II

DOPAMINE

↓ II

DESIRED RENAL VASODILATION

Fig. 8.2 I: L-glutamyl transpeptidase

II: L-aminoacid decarboxylase.

Liver targeting

Site selective transport pathway in liver for bile acid transport is associated with sinusoidal membrane of the hepatocytes. E.g. Chlorambucil.

Site directed prodrugs

Bioreversible prodrug is carried to a site of action due to increased concentration at the target cells, bioactivation occurs.site directed prodrugs reqiures a carrier and drug. These prodrugs concentrate on macromolecular carrier. Polymeric carrier can be divided into stable forms in the body and undergoes degradation to small fragmentation *in vivo*. Nonbiodegradable polymers contain C-C, ether stable bonds. Biodegradable or non accumulated macromolecules suitable for systemic use include polypeptide, polysaccharides, macromolecular carriers.

Use of Prodrugs in Novel Drug System

The novel drug systems improves the therapeutic effectiveness of a drug by providing sustained, controlled delivery and targeting the drug to the desired site. The effectiveness of a system depends on the entrapment efficacy, which in turn affected by physicochemical properties of drug. Such properties can be modified by the prodrugs, therefore prodrugs have been developed and utilized successfully. Prodrugs are synthesized mostly as ethers, amides forms where -OH, -COOH, NH_2 free groups of drug covalently binded with promities. Prodrugs must be hydrolyzed in the body by esterases, amidase.

Some Strategies of Prodrugs

1. Prodrugs incorporated into a delivery system

Colloidal systems of novel drug delivery produces sustained and controlled drug delivery by releasing the encapsulated drug, when large amount of carrier is needed for encapsulating drug a low dose of drug, then that drug delivery system is neither economic nor administrable. Low encapsulation efficacy also produces unsatisfactory results. So, 100% encapsulation efficacy is required for an ideal drug delivery system. Prodrug can be formulated as,

1. Liposomes
2. Neosomes

3. Emulsons
4. Solid lipid nanoparticles
5. Lipoparticles

Prodrug in Liposome Forms

The term liposome was derived on the basis of names of sub cellular particles like ribosomes, lysosomes. Liposomes are the vesicles consists of lipid bilayer particles enclosing an aqueous environment. The lipid which is the most commonly used in the liposomal preparation is phospholipid like sphingolipid, glycolipid, sterols. As liposomes possess both hydrophilic region and lipophilic region both hydrophilic, lipophilic drugs are incorporated into liposomes. Examples is 5-fluorouridine is an anticancer drug, which shows encapsulation efficacy and rapid leakage during storage. Where as its lipophilic prodrug showed high encapsulation efficacy and its palmitoyl prodrug did not show any drug leakage during storage, no aggregetation of liposomes, no precipitation of drug, no change in vesicle morphology.

Prodrugs in Niosomes Formulaton

Niosomes are inexpensive alternative to liposomes. These are closed vesicles formed from nonionic surfactant in aqueous media. Nonionic surfactant used for the preparation of niosomes are cheap, chemically stable.

Prodrugs in Emusions

In the sustained drug delivery, passive targeting to macrophages and in active targeting by ligand attachment oil in water emulsions are used. Etoposide and teniposide derivatives of podophyllotoxin are poorly soluble in lipid emulsion with cholesterol ester as oil component. Its esterified phenolic-4'-hydroxyl group of drug with unsaturated fatty acids forms lipophilic prodrugs, which are soluble in lipid emulsion. These emulsion showed good physical stability and no leakage of entrapped drug during storage.

Prodrugs in Solid Lipid Nanoparticles

Solid lipid nanoparticles comprises of solid core triglyceride and a coat of phospholipid. Solid lipid nanoparticles show combined advantages of both fat emulsions and liposomes. Natural lipids are used in solid lipid nanoparticles and incorporation of in triglyceride core shows prolonged effect. Incorporation of azidothymin on solid lipid nanoparticles was low but its ester prodrugs i.e., Azidothymin-Palmiyaye ester incorporated well up to 90%.

Pharmaceutical Applications

1. Improvement of Taste

Many drugs show bitterness which is the reason for poor patient compliance. One of the approach to mask the bad taste of the drugs is decrease the drug solubility in the saliva so that it bioavailability decreases results in lower the affinity of drug towards taste receptors, thus bitterness of a drug can be overcome.

Parent drug	Taste	Prodrug with improved taste
Chloramphenicol	Bitter	Palmitate ester
Triamcinolone	Unpleasant	Diacetate ester

2. Improvement of Odour

Vapour pressure affects the odor of many drugs. A drug with high vapor pressure show strong odor. E.g., Ethyl mercaptan, anti-leprotic drug, is a foul smelling liquid of boiling point 35°C. It is converted to its phthalate ester form which is odorless.

3. Change in Physical form of A Drug

Some drugs in liquid form are not suitable for use in the formulation of the tablets when the is high. That liquids can be converted into the solid prodrugs which involves the formation of the symmetric molecules having greater tendency to crystallize. E.g. Ethyl ester of mercaptan and tricholroethanol.

4. Reduction of Pain after Injection

When the drugs precipitates or penetrates into the surrounding cells or when the solution is strongly acidic, alkaline or alcoholic are painful on intra muscular injections. E.g., Acidic form of clindamycin i.e.,clindamycin hydrochloride shows low aqueous solubility so its ester form is preferred. Its ester form is the 2-phosphate ester of clindamycin.

5. Reduction of Git Ittritation

Several drugs cause irritation and damage of the gastric mucosa by increasing the acid secretion or by interacting with the protective mucosa.

Parent drug	Prodrug that cause little irritation
Salicylic acid	Salsalate, aspirin
Oleandrin	Oleandrin acetate

6. Enhancement of Solubility and Dissolution Rate of Drug

For hydrophilic drugs dissolution is the rate limiting step in the absorption so hydrophilic drugs with hydroxyl functional groups can be converted to hemi-esters, hemi-glutarates or hemi-phthalates drugs with phenolic groups like cortisol, prednisolone, betamethasone which has poor water solubility, phosphate ester are preferred.

7. Enhancement of Chemical Stability

Azacytidine, antineoplastic drug, its aqueous solution is readily hydrolyzed but its bisulphate prodrug is stable to degradation at acidic pH than the parent drug.

8. Enhancement of Bioavailability

Dipalmitoyl ester of naproxen produces less irritation and high plasma concentration, so bioavailability is increased.

CHAPTER 9

RESEALED ERYTHROCYTES

Introduction

Erythrocytes are one of the potential carrier systems in the human body have been thoroughly explored in recent times for targeted drug delivery among various formed elements of blood such as WBC, platelets etc. The RBC has been exploited as novel cellular carriers due to their intrinsic carrier physiology, and further more they are natural and possibly devoid of untoward immunological insults.

Criteria for Delivery of Drug through RBC

- The drugs that are being employed for entrapment must permeate through the RBC membrane without any alteration in structure, composition and function of the membrane.
- The carrier cell must release the drug in a desired manner upon reaching the desired target site.
- The entrapment process should be carried by inducing a reversible, transient, and permeability change.
- The bioactive components meant for incorporation may be any drug molecule, enzyme or DNA.

Sources of Erythrocytes for Drug Loading, Releasing and Drug Delivery

- Mammalian, mice, pigs, dogs, sheep, goats, monkeys, rats and rabbit erythrocytes have been studied for drug loading.
- Collection of blood in mammalians carried through vein puncture and in rodents it is through orbital sinus or cardiac puncture.
- Fresh animal blood is employed. Fresh whole animal blood means any blood collected and chilled immediately to 4°C and stored for not more than 2 days.

- The centrifugal force required for sedimentation and composition of buffer meant for washing RBC varies according to the species from which blood is collected.

General Procedure

Blood is collected into heparinized tubes subjected for centrifugation at 1000 g for 5 minutes. Packed RBC thus obtained is separated from plasma. Washing of packed RBC with buffer solution (NaCl 140 mmol/l) is carried followed by storing of RBC in acid-citrate-dextrose buffer at 4°C.

Methods for Drug Loading

In general hypotonic lysis of cells in a solution containing the drug/enzyme to be entrapped followed by restoration of tonicity to reseal them. This serves as the basic principle in drug a loading procedure. But other techniques such as electrical breakdown, endocytosis, and chemical perturbation of membrane, lipid and erythrocyte fusion have also been explored. Some of the procedures are as follows.

Hypotonic Haemolysis and Isotonic Resealing Methods

Four variations of the procedure have been described based upon which hypotonicity is produced. The ghost populations so obtained are heterogeneous in nature. Three types of ghosts can be described.

Type I ghosts - these reseal immediately after haemolysis.

Type II ghosts - these reseal after reversal of haemolysis.

Type III ghosts - these remain leaky under different experimental conditions.

The ratio of three fractions depends upon temperature at which haemolysis is affected and on time interval between haemolysis and restoration of tonicity. Erythrocytes have an exceptional capability for reversible shape changes with or without accompanying volume changes. Increase in volume initially leads to conversion of normal biconcave discocytes (normal erythrocytes) to spherocytes. This is because they do not have superfluous membrane or internal membranes and capacity to synthesize additional plasma membranes due to the absence of nucleus in mature RBC; hence the surface area is inevitably fixed. Thus the cells become spheres as they accommodate additional volume with fixed surface area. However, these swollen erythrocytes have little capacity to resist volume greater than 50-75% of the initial volume and when placed in solution less than about 150 mOsm/kg (corresponding to about 0.4% w/v NaCl), the membrane ruptures, permitting escape of cellular components. Erythrocytes are resealed on addition of sufficient 1.54 M KCl, which restores isotonicity.

Dilutional haemolysis

In this, erythrocytes when exposed to hypotonic saline solution (0.4%) swell until it reaches a critical value of volume or pressure where the membrane of RBC's ruptures and becomes permeable to macromolecules and ions. This permits the escape of cellular components.

Preswell dilutional haemolysis

In this technique, initially controlled swelling of erythrocytes is seen which is followed by centrifugation at low "g" to take them up to point of lysis. After this, a small volume of drug solution is added to obtain drug loaded resealed erythrocytes. Haemolysis and resealing is implemented in three steps:

In first step, one volume of washed cells is suspended in five volumes of slightly hypotonic buffer (0.6% w/v NaCl). Under these conditions, only a small percentage of cells will be lysed and will constitute the most fragile population of erythrocytes.

Second, after incubation at 0°C for 5 minutes and the cells are recovered by gentle centrifugation.

Third, the swelling procedure is further extended by addition of a volume of hypotonic buffer equal to one half the volumes of the swollen cells to effect lysis of the cells. This hypotonic buffer medium also contains the material to be loaded. The cells are allowed to remain lysed for 10 minutes at 0°C. This is followed by restoration of tonicity and resealing of the membrane. Gentle swelling of cells results in good retention of cytoplasmic constituents thus a good survival *in vivo*, having entrapment efficiency as high as 72%.

Isotonic osmotic lysis

This method is proposed to over come the potential disadvantages of hypotonic haemolysis of cell. Haemolysis in isotonic solutions can be achieved both by chemical and physical means. If the erythrocytes are incubated in solutions of a substance with high transerythrocytic membrane permeability, the solute will diffuse into the cells. This is mainly due to inwardly directed chemical potential gradient. This will be further followed by water uptake which continues until osmotic equilibrium is restored. Variation in this method include:

Polyethylene glycol induced haemolysis: In this technique transient permeability is produced in the erythrocyte membrane using polyethylene glycol and thus establishing equilibrium within the environment with resultant diffusion of bioactive drugs.

Ammonium chloride induced haemolysis: Ghost cells obtained by ammonium chloride induced haemolysis exhibit properties very similar to normal cells with respect to their permeability to sulfates, monosaccharides, sucrose and L-lactate (small molecules). However, we can not to rule out changes in the membrane structure and composition due to the presence of membrane lytic substances.

Dialysis

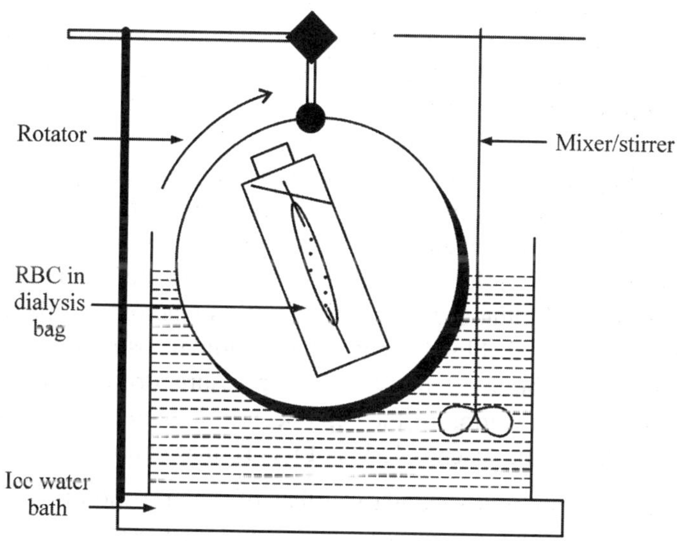

Fig. 9.1 Dialysis.

The low entrapment efficiency of the dilution procedure can be overcome by carrying out lysis and resealing within a dialysis tube. The common principle in this is semi permeable dialysis membrane that maximizes the intracellular: extracellular volume ratio for macromolecules during lysis and resealing, and also allow for free flow of small ions haemolysis and resealing is carried in three steps.

First, washed erythrocytes with value haematocrit 85- 95% are mixed with phosphate buffered saline (0.15 M NaCl, 5 mM sodium phosphate, pH 7.4) containing the drug protein to be entrapped. The dialysis bag is inflated with an air bubble which is critical and sealed in such a way that the erythrocyte suspension occupies not more than 75% of the internal volume. During lysis and resealing, it traverses through the length of the dialysis tubing and serves to blend the contents of the dialysis bag.

The sealed dialysis bag is then placed in a bottle containing at least 200 ml of lysis buffer (0.1% w/v NaCl, 0°C) and placed on a mechanical rotator for 2 hours at 4°C. After the initial lysis at 0°C, the dialysis bag is transferred to a bottle containing at least 200 ml of resealing buffer at room temperature for 30 minutes.

Electro-insertion or electro-encapsulation

The main principle involved in this method is creating electrically induced permeability changes at high membrane potential differences. Electric breakdown is evident when membrane is polarized for microseconds using varied voltage values. The extent of pore formation depends upon the electric field strength, pulse duration and ionic strength of suspending medium. Once membrane is perforated, regardless of the size of pores, ions rapidly distribute between the extra and intracellular space, however the membrane still remains impermeable to its cytoplasmic macromolecules. The cell membrane eventually lyses because of the colloidal osmotic pressure of its macromolecular contents. In the case of red blood cells, the colloidal osmotic pressure of haemoglobin is about 30 mOsm. This pressure drives water and ion influx, as a result swelling of the cells occurs. The membrane is ruptured when the cell volume reaches 155% of its original volume.

Thus, cell lysis is a secondary effect of electric modification of the membrane. Since the cell lysis is due to colloidal osmotic swelling, the rational to prevent lysis is to balance the colloidal osmotic pressure of cellular macromolecules. This can be affected by addition of large molecules (like tetra saccharide stachyose or protein such as bovine serum albumin) and ribonucleases. This helps to counteract the colloidal osmotic swelling of electrically perforated erythrocytes. Under this osmotically balanced condition pores stay open at 4°C for few days. If drug molecules are added at this point, they permeate into RBCs. A suitable procedure could be subsequently used to reseal these pores.

Entrapment by endocytosis

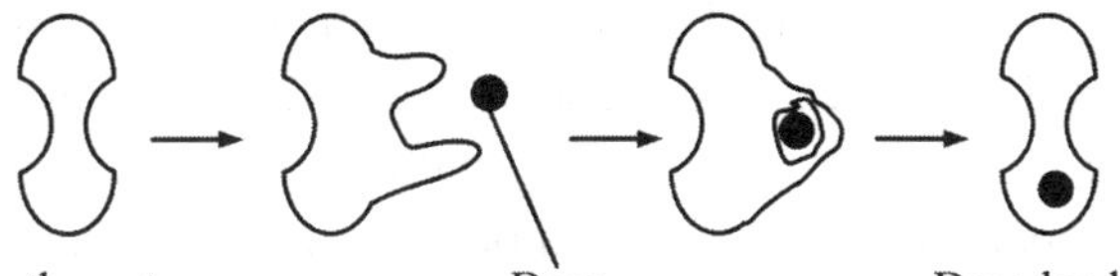

Fig. 9.2 Entrapment by Endocytosis.

Intracellular vesicles can be induced in erythrocytes containing small molecules, drugs or viruses from external medium. In this method, the vesicle membrane separates the endocytosed substance from the cytoplasm containing drug, which is sensitive to inactivation by cytoplasmic enzymes and also protects the erythrocyte membrane. The resulting erythrocytes contain vacuoles and probably have different *in vivo* survival characteristics from resealed cells, prepared by using other methods. The swollen ghosts so prepared exhibit larger endocytic vacuoles. The drug substances are entrapped in these endocytic vacuoles. Drug induced endocytosis is quite common and a variety of ampiphilic cations/drugs produce first stomatocytosis and then mostly at the advancing lip of the stoma, forms inside-out endocytic vacuole. This method is efficient for loading large particles such as viruses, enzymes and small molecules.

Membrane perturbation

The principle involved in this method is increasing the permeability of the erythrocytic membrane when exposed to some chemical agents. This increase in permeability allows small molecular weight substances to get entrapped.

Lipid fusion method

In this the exchange of lipid entrapped drug is done by directly fusing the lipid vesicles containing drug can be directly fused with human erythrocytes. This method suffers from very low encapsulation efficiency.

In vitro characterization

Resealed erythrocytes after loading are characterized for *in vitro* tests. These characterizations are vital to ensure their *in vivo* performance and therapeutic benefits. Some of the characterization methods include

Drug content

In this method, packed loaded erythrocytes are first deproteinised with acetonitrile and then subjected to centrifugation at 2500 rpm for duration of 10 minutes. The clear supernatant matter is analyzed for the drug content using specified estimation methodology for entrapped drug.

In vitro drug and hemoglobin release

In this method, normal and loaded erythrocytes are incubated at 37 ± 2°C in phosphate buffer saline PBS (pH 7.4) at 50% haematocrit in a metabolic rotating wheel incubator bath. Periodically, the samples are withdrawn from the medium by using a hypodermic syringe fitted with a 0.8 μ spectropore membrane filter. The samples are then deproteinised with acetonitrile and can be estimated for the amount of drug released. Percent hemoglobin can similarly be calculated at various time intervals at 540nm spectrophotometrically.

Osmotic fragility

It is the parameter, which simulates and mimics the bioenvironmental conditions that are encountered on *in vivo* administration, *in vitro* handling and the effect of loaded contents on the survival rates of the erythrocytes.

When red blood cells are exposed to solutions of varying tonicities, it results in the change of their shape due to osmotic imbalance. To evaluate the effects of varying tonicities on RBCs, drug loaded erythrocytes are incubated with saline solutions of different tonicities at 37±2°C for 10 minutes. These suspensions after centrifugation for 15 minutes are assayed for drug or haemoglobin release, which should be in the acceptable range for system to be therapeutically effective.

Osmotic shock

This parameter evaluates the ability of resealed erythrocytes to withstand the stress and maintain their integrity as well as appearance. The procedure involves the Incubation of the resealed erythrocytes with distilled water for a duration of 15 minutes which is followed by centrifugation at 3000 rpm for 15 minutes. This leads to the release of haemoglobin to varying degrees, which could be estimated spectrophotometrically.

Turbulence shock

Resealing of the erythrocytes makes them sensitive towards turbulence or mechanical agitation and hence an estimation of turbulence shock study provides their expected performance *in vivo*. This parameter evaluates the effects of shear force and pressure by which resealed erythrocyte formulations are injected.

Percent cellular recovery

The shape, size and surface features of loaded erythrocytes are evaluated by the microscopic methods like Phase contrast optical microscopy, transmission electron microscopy and scanning electron microscopy. Percent cell recovery can be determined by measuring the number of intact erythrocytes that remain per cubic mm. the measurement of these erythrocytes is done with the help of haemocytometer.

Mechanism of release of resealed erythrocytes

- Phagocytosis
- Diffusion through the membrane of cell
- Using a specific transport system

The rate of diffusion depends upon the rate at which a particular molecule penetrates through a lipid bilayer. Many substances enter cells by a specific membrane protein system because the carriers are proteins with many properties analogous to that of enzymes, including specificity e.g. nucleotides and nucleosides. Release of drugs from erythrocytes is rapid followed by sustained release profile and rate of exit is proportional to the instantaneous intracellular drug concentration which follows first order kinetics. By incorporating polymer into erythrocytes, the release pattern may be modified.

Shelf and storage stability of resealed erythrocytes

Storage of resealed erythrocytes poses a major challenge in their practical utility as drug delivery system cryopreservation of erythrocytes in liquid nitrogen is mostly utilized method for storage. The most common storage media include Hank's balanced salt solution and acid-citrate-dextrose at 4°C by addition of calcium-chelating agents.

In vivo survival and immunological consequences

Erythrocyte loading using hypotonic haemolysis followed by isotonic resealing; dialysis and electro encapsulation methods appears to sustain normally in circulation. A rapid loss of cells during first 24 hours followed by much slower loss afterwards. The early loss possibly accounts for removal of nearly 15% of total cell population; this represents the cells that are severely damaged during the loading procedures. The second phase has half-life of orders of weeks for different mammalian erythrocytes. It is observed that erythrocyte carriers constructed of autologous sources do not elicit immunological consequences. However, it is customary to realize that during loading process some antigenic impurities may get entrapped resulting in immunological manifestations. The erythrocytic carriers constructed of red blood cells of mice, cattle, pigs, dogs, sheep, goats, and monkeys exhibit a comparable circulation profile with that of normal unloaded erythrocytes. On the other hand, resealed erythrocytes prepared from red blood cells of rabbits, chickens, and rats exhibit relatively poor circulation profile.

Biomedical Applications of Resealed Erythrocytes

Erythrocytes as Carriers for Enzymes

The first report of successful clinical trials of the resealed erythrocytes loaded with enzymes for replacement therapy is that of ß-glucoserebrosidase for the treatment of Gaucher's disease. The disease is characterized by inborn deficiency of lysosomal ß-glucoserebrosidase in cells of RES thereby leading to accumulation of ß-glucoserebrosides in macrophages of the RES.

The most important application of resealed erythrocytes in enzyme therapy is that of asparginase loading for the treatment of pediatric neoplasms. This enzyme degrades aspargine, an amino acid vital for cells. This treatment prevents remission of pediatric acute lymphocytic leukemia. Other enzymes used for loading resealed erythrocytes include urease, galactose-1-m phosphate uridyl transferase, uricase and acetaldehyde dehydrogenase.

Erythrocytes for Drug Targeting

Osmotically loaded erythrocytes can act as a drug carrier in systemic circulation. Chemically surface modified erythrocytes are targeted to organs of mononuclear phagocytic system because of changes incorporated in the membrane that are recognized by macrophage cells.

Drug Targeting to Reticuloendothelial System

Though, resealed erythrocytes have been proposed for passive auto-vectorization to reticuloendothelial system where modified surface characteristics impart them the selectivity and specificity towards target cells. Various approaches for modifying the circulatory and tissue uptake parameters of resealed erythrocytes to make them site specific and target oriented are adopted.

1. Surface modification with antibody

Circulatory half life is shortened on coating resealed erythrocytes with antibodies. Lightly coated resealed erythrocytes cause a reduction of half life from 27 days to several minutes and majority of erythrocytes are sequestered by spleen and liver macrophages and as a result get accumulated in these organs.

2. Surface modification involving carbohydrates

Enzymatically modified cell surface carbohydrates of erythrocytes exhibit a different biodistribution as compared to their plain version. Sialidase, an enzyme that removes sialic acid from external glycoprotein and glycolipids causes significant reduction in the circulatory half life of erythrocytes and make them prone to macrophage uptake especially in liver and spleen.

3. Surface modification with sulphydryls

External sulphydryls of erythrocytes can be oxidized to varying degrees and thus circulatory half life can be manipulated from a normal value of 27 days to several minutes.

Drug Targeting to Liver

1. Treatment of liver tumors

Anticancer agents encapsulated in erythrocytes can be used for targeting hepatic carcinomas. E.g., Bleomycin, adriamycin, carboplatin, gentamycin, and methotrexate, daunorubicin.

2. Treatment of parasitic diseases

Resealed erythrocytes are versatile drug carriers for the delivery of antiparasitic agents. Parasitic diseases in which the parasite resides in reticuloendothelial system have been effectively treated with resealed erythrocytes.

E.g. Pentamidine, primaquine phosphate and metronidazole.

3. Removal of RES iron overload

In iron overload resulting from repeated blood transfusion, reticuloendothelial cells are the primary and the major sites of iron accumulation. Iron chelating drug (desferrioxime) has been entrapped in erythrocytes, for promoting excretion of iron that is present as intracellular ferritin and haemosiderin deposit. The drug forms soluble chelates and depletes the depots.

4. Removal of toxic agents

Resealed erythrocytes containing a recombinant phosphotriesterase which hydrolyze many organophosphorus compounds, including paraoxon, a potent cholinesterase inhibitor to antagonize organophosphorus intoxication.

Oxygen Deficiency Therapy

Resealed erythrocytes are also used in hypoxic where an improved oxygen supply is required as in the following cases:

- High altitude conditions.
- Small number of alveoli.
- Increased resistance to oxygen diffusion in the lungs.
- Reduction in oxygen transport capacity.
- Liver mediated detoxification processes.ss

Delivery of Macromolecules

Biological functions of macromolecules such as DNA, RNA, and proteins are exploited for various cell biological applications. Hence, various methods are used to entrap these macromolecules into cultured cells (e.g., microinjection). In microinjection, erythrocytes are used as micro syringes for injection to the host cells.

Conclusion

Until other carrier systems come of age, resealed erythrocytes technology will remain an active arena for the further research. In near future, erythrocytes based delivery system with their ability to provide controlled and site-specific drug delivery will revolutionize disease management. The use of resealed erythrocytes looks promising for a safe and effective delivery of various drugs for passive and active targeting. However, the concept needs further optimization to become an outline drug delivery system. The same concept can also be extended to the delivery of biopharmaceuticals and much remains to be explored regarding the potential of resealed erythrocytes.

CHAPTER 10

TRANSDERMAL DRUG DELIVERY SYSTEMS

Introduction

Transdermal drug delivery (TDDS) is the non invasive delivery of medications from the surface of skin. Skin is the largest and most accessible organ of human body through its layers to the circulatory systems.

TDDS has many advantages over conventional injection and oral methods.some of them are listed here:

1. It enhances patient compliance.
2. Minimizes harmful side effects of drug caused from over dose.
3. It is more convenience especially notable in patches that require only once weekly application.
4. Such a simple dosing regimen can aid in patient adherence to drug therapy. The development of TDDS is multi-disciplinary activity that encompasses fundamental.

Definition

The transdermal patch is a medicated adhesive pad that is design to release the active ingredient at a constant rate over a period of several hours to days after application to the skin. It is also called skin patch. A skin patch uses a special membrane to contained within the patch can pass through the skin and into the blood stream.

Advantages

1. Avoidance of first pass affect.
2. Avoid GIT difficulties which are caused by enzymes and food interactions.
3. Stable and controlled blood level.
4. Ease of termination of drug action if necessary.

5. Long duration of action.
6. Less chances for over dose.
7. No interference with gastric and intestinal fluids.
8. Comparable characteristics with intravenous infusion.
9. Suitable for administration of drugs having.
10. Very short half life ex: Nitro glycerin
11. Narrow therapeutic window
12. Poor oral bioavailability.
13. Maximum flexibility and gentleness during use.
14. Immediate, visible effects.
15. This route is suitable for other than oral route because incase of

Disadvantages

1. The transdermal delivery is unsuitable when the drug dose is large.
2. Drug has large molecular size makes absorption difficult should ideally below 800 to 1000 Daltons.
3. Drug is skin sensitizing and irritating.
4. Drug is metabolized in skin.
5. Drug undergoes protein binding in skin.
6. Drug is highly lipophilic or hydrophilic.
7. Some patients may develop contact dermatitis at the site of application.

Anatomy of Skin

The skin is a major layer of our body. It has mainly epidermis and dermis. The top layer is called the epidermis. The Stratum Corneum (SC) is the outer most layer of the epidermis which is responsible for the barrier property of the skin. SC layer comprises multiple layers of flattened epithelial cells. More than 90% of all cells in the SC or corneocytes. These cells are keratinized and the cells are terminally differentiated and tightly packed. The internal space in the SC is occupied by lipids. SC lipids are generally non polar and enriched with free fatty acids, free sterols with smaller quantities of glycolipids, sterols esters and triglycerides, cholesterol sulfate and hydrocarbons are present as well. The lipids are arranged in the form of lamellar stacks with the polar head groups of lipids gathered in layers while the non polar chains point in opposite directions.

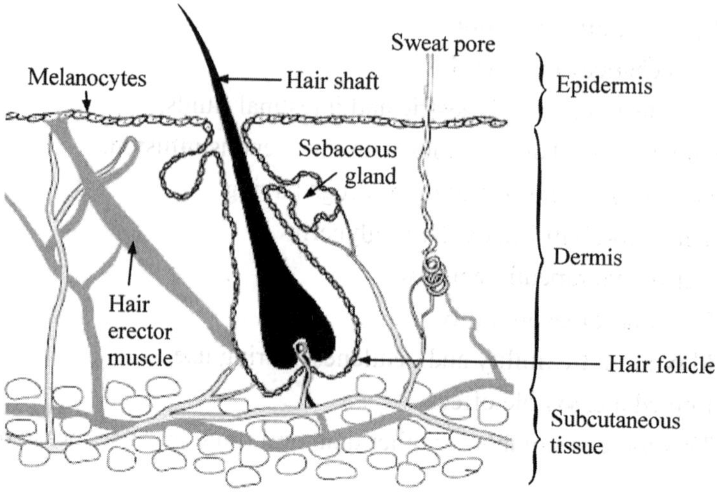

Fig. 10.1 Skin cross section.

The cells beneath the SC from the viable epidermis. This is physicochemically similar to other living tissues. The cells have thin membranes that encapsulate cytoplasm and are held together by ton fibrils. Melanocytes which are not very abundant are the cells in the epidermis that produced the pigment melanin. Melanin absorbs harmful UV light before UV radiation can react the nucleus.

Just beneath the epidermis is the dermis. The dermis consists mostly of connective tissue and is much thicker than the epidermis. It is responsible for the skin pliability and mechanical resistance and is also involved in the regulation of body temperature. The dermis supplies the vascular epidermis with nutrients by means of its vascular network. It contains the sense organs for touch, pressure, pain and temperature as well as blood vessels, nerve fibers, sebaceous and sweat glands and hair follicles. The dermis has a dense network of capillaries. Thus it is considered to be the primary site of material exchange between the skin and blood.

Transdermal Permeation

Skin is considered as an impermeable protective barrier of the body but later investigations were carried out which provide the utility of skin as a route for systemic administration of drugs. Skin is the most intensive and readily accessible organ of the body. This is so said as only a fraction of millimeter of tissue separates its surface from the underlying capillary network. The various steps involved in transport of drug from patch to systemic circulation are as follows

Diffusion of drug form the drug reservoir

↓

Rate controlling membrane

↓

Stratum corneum

↓

Sorption of the drug by stratum corneum

↓

Sorption of the drug by stratum corneum

↓

Uptake of drug by capillary network in the dermal papillary layer

↓

Effect on target prgan

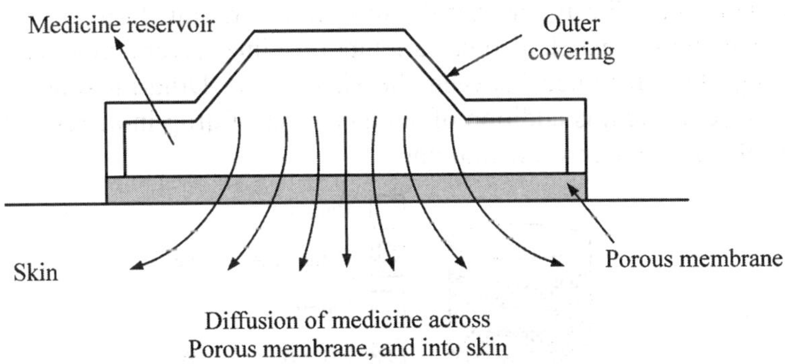

Fig. 10.2 Transdermal patch.

Types of Transdermal Patches

There are four main types of transdermal patches:

(i) Single layer drug in adhesive

(ii) Multi layer drug in adhesive

(iii) Drug reservoir in adhesive

(iv) Drug matrix in adhesive

Single-layer drug-in-adhesive

In this system, the formulated drug is included directly within the skin-contacting adhesive. In this type of patch, the adhesive layer is responsible for the release of the drug from the formulation, and this serves to adhere the various layers together, along with the entire system to the skin. The adhesive layer is surrounded by a temporary liner and a backing.

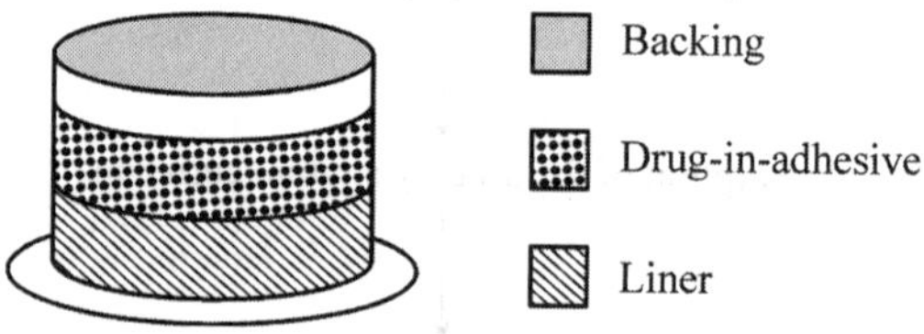

Fig. 10.3 Single-Layer Drug-in-Adhesive.

Multi-layer drug-in-adhesive

The similarity between the Multi-layer Drug-in-Adhesive and Single-layer Drug-in-Adhesive in that the drug is incorporated directly into the adhesive and this patch also has a temporary liner-layer and a permanent backing. The differnce between the single-layer-Drug-in-Adhesive and multi-layer system is addition of another layer of drug-in-adhesive which is usually separated by a membrane.

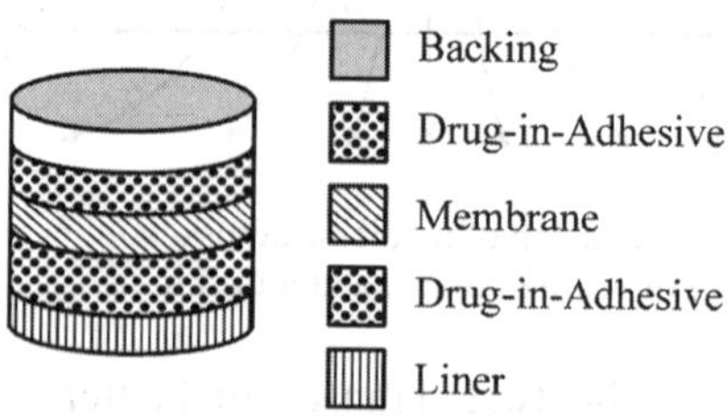

Fig. 10.4 Multi-Layer Drug-in-Adhesive.

Drug reservoir-in-adhesive

The reservoir transdermal system design includes a liquid compartment containing a drug solution or suspension which is separated from the release liner by a semi-permeable membrane and adhesive. Here, the adhesive component of the product can be as a continuous layer between the membrane and release liner or as a concentric configuration around the membrane.

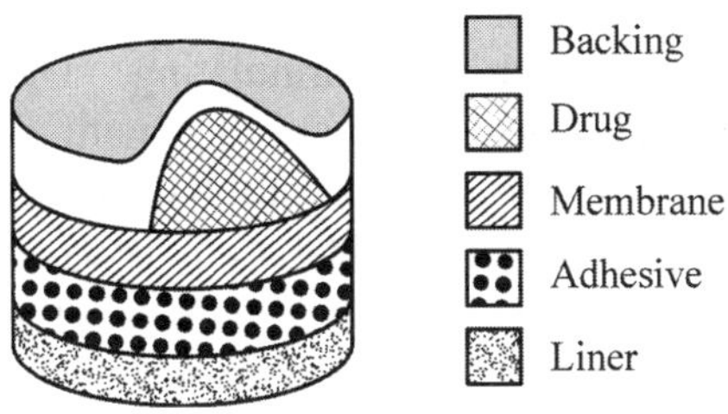

Fig. 10.5 Drug Reservior-in-Adhesive.

Drug matrix-in-adhesive

In this, the matrix system has a drug layer of a semisolid matrix which contains drug either as a drug solution or suspension, which is in direct contact with the release liner. The drug layer is surrounded by the adhesive layer in this patch.

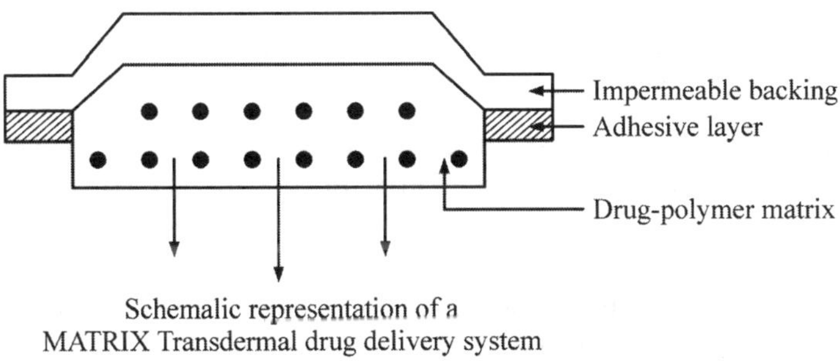

Fig. 10.6 Drug Matrix in-Adhesive.

Basic Components of TDDs

The basic components of TDDS are listed here under:

- Polymer matrix / Drug reservoir
- Drug
- Permeation enhancers
- Pressure sensitive adhesive(PSA)
- Backing laminates
- Release liner
- Other excipients like plasticizers and solvents.

Polymer matrix/drug reservoir

The polymers form the back bone of TDDS. These polymers control the drug release from the device. Polymer matrix can be prepared by drug dispersion in either liquid or solid state synthetic polymer base. The polymers which are used in formulating TDDS should have biocompatibility and chemical compatibility with the drug and other components of the system such as penetration enhancers and PSAs. They should also provide consistent and effective delivery of a drug throughout the products intended self life. The polymers should not cause any harm to the skin.

The polymers used for TDDS can be classified as follows based on their origin:

- ***Natural Polymers:*** E.g., Cellulose derivatives, gelatin, shellac, waxes, gums etc.
- ***Synthetic Elastomers:*** E.g., Polybutadiene, hydrin rubber, polyisobutylene, silicon rubber, nitrile, acrylonitrile, neoprene, butylrubber etc.
- ***Synthetic Polymers:*** E.g., Polyvinyle alchohol, polyvinyl chloride, polyethylene, polypropylene, polyacrylate, polyurea, polyvinylepyrrolidone, polymethylemethacrylate etc.

Other polymers like cross linked polyethylene glycol, ethyl cellulose, poly vinyl pyrrolidine are used as matrix formers for TDDS.

Drug

The transdermal route is advantageous route for the drugs with appropriate pharmacology and physical chemistry.

- Transdermal patches are mostly used for drugs which undergo extensive first pass metabolism.
- Drugs with narrow therapeutic window which causes noncompliance due to frequent dosage.

The foremost requirement of TDDS is that the drug must possess the right mixes of both physicochemical and biological properties for transdermal drug delivery. It is generally accepted that the best drugs for passive adhesive transdermal patches should be non ionic of low molecular weight have adequate solubility in oil and water. A low melting point and drug must be more potent. In addition drugs like rivastigmine used for Alzheimer and Parkinson dementia and rotigotine used for Parkinson, methylphenidate used for attention deficit hyperactive disorder.

Permeation enhancers

These are the chemical compounds which increase permeability of stratum corneum so as to attain higher therapeutic levels of drug candidate. Penetration enhancers interact with structural components of stratum corneum. It is proteins or lipids which alter the protein and lipid packaging of stratum corneum which thus chemically modify the barrier functions ultimately leading to increased permeability.

Pressure sensitive adhesives

PSA is a material which helps in maintaining an intimate contact between transdermal system and the skin surface. They have properties like

- It should adhere with not more than applied fingure pressure
- Be aggrisevily and permanently tachy.
- Exert a strong holdingforce
- It should be removable from the smooth surface with out leaving a residue

Some examples are polyacrylates, polyisobutylene and silicone based adhesives are widely used in TDDSs. The selection of an adhesive is based on numerous factors along with the patch design and drug formulation.

For matrix systems with a peripheral adhesive, an incidental contact between the adhesive and the drug.

Penetration enhancers should not disturb the stability of drug. In case of reservoir systems which include a phase adhesive, the drug that diffuses must not affect the adhesive. In case of drug in adhesive matrix systems, the selection is based on the rate at which the drug diffuses or penetrates and the rate at which penetration enhancers will diffuse through the adhesive. Ideally PSA should be physicochemically and biologically compatible and should not alter the release of the drug.

Backing Laminates

While designing a backing layer, one must consider the chemical resistance of material as it is most important. Compatibility with the excipient should also be considered because the contact between backing layer and the excipient may lead to the leaching of the additive from the backing layer or may cause excipients to diffuse. However an overemphasis on the chemical resistance is not good because it may lead to stiffness and high occlusivity to moisture vapor and air causing patches to lift and possibly irritate the skin during long wear. The most comfortable backing layer is the one that exhibits lowest modular or high flexibility, good oxygen transmission and a high moisture vapor transmission rate.

E.g., Vinyl, polyethylene and polyster films.

Release liner

During storage the patch is covered by a protective line which is the release liner that is removed and discharged immediately before the application of the patch to skin. It is therefore considered as a part of the primary packiging material.The liner is in intimate contact with the delivery system, so it must be chemical inert to the drug and other excipients used like penetration enhancer and water. The composition of the release liner is a base layer which may be either non occlusive (E.g., paper fabric) or occlusive (E.g., polyetylene,polyvinyl chloride) and a release coating layer which is made up of silicon or teflon. Other materials of release liner include foil made of polyster and metalised laminates.

Other excipients

The drug reservoir is made up of various solvents such as chloroform, methanol, acetone, isopropanone and dichloromethane. In addition, plasticizers like dibutyl pthallate, triethylcitrate, polyethylene glycol and propylene glycol are also added to provide plasticity to the transdermal patches.

Preparation of Different Types of Transdermalpatches

Various systems are being introduced in market. These can be classified into following types:

- Matrix type
- Reservoir type
- Membrane matrix hybrid
- Micro reservoir type
- Drug in adhesive type

Matrix Type Transdermal Patchs

In this, drug reservoir is prepared by dissolving the drug and polymer in a common solvent. The insoluble drug must be homogenously dispersed in hydrophilic or lipophilic polymer. Then plasticizer like dibutyl pthalate, triethylcitrate, poly ethylene glycol or propylene glycol and permeation enhancer are added in required quantity and mixed properly. This forms the medicated polymer which is then molded into rings with a specific surface area and controlled thickness over the mercury on horizontal surface followed by solvent evaporation at an elevated temperature. The film thus formed is then separated from the rings, and then mounted on to an occlusive base plate in a compartment fabricated from a drug impermeable backing. Adhesive polymer formed is then spread along the circumference of the film. Examples of some commonly used polymers for matrix are cross linked polyethylene glycol, ethyl cellulose, poly vinyl pyrrolidone and hydroxy propyl methyl cellulose.

The dispersion of drug particles in the polymer matrix can be acheived by either homogenously mixing the finely ground drug particles with a liquid polymer or highly viscous base polymer followed by cross linking of polymer chains or homogenously blending drug solids with a rubbery polymer at an elevated temperature. The matrix system is exemplified by the development of Nitro-Dur. Advantages of matrix patches are listed below:

- Absence of dose dumping,
- Direct exposure of polymer matrix to the skin and
- No interference of adhesive.

Reservior Type Transdermal Patch

The drug reservoir is made up of a homogenous dispersion of drug particles suspended in an unreachable viscous liquid medium (E.g., silicon fluids) in order to form a paste like suspension or gel or a clear solution of drug in a releasable solvent (E.g., ethanol).The drug reservoir thus formed is sandwiched between a rate controlling membrane and backing laminate.

The rate controlling membrane can be of non porous nature so that the drug is released by diffusing directly through the material. The material may contain fluid filled micropores in which case the drug may additionally diffuse through the fluid, thus filling the pores. In the case of nonporous membrane, the rate of passage of drug molecules generally depends upon the solubility of drug in the membrane and the thickness of membrane. Hence the choice of membrane material depends on the type of drug being used. The dose rate per unit area of the device can be controlled by varying the composition and thickness of the membrane. EVA, ethyl cellulose, silicon rubber and polyethanes are commonly used to prepare rate controlling membranes. EVA is used mostly to prepare rate controlling membrane in transdermal delivery system because it allows the membrane permeability to be altered by adjusting the content of vinyl acetate polymer. Polyurethane membranes are suitable mainly for hydrophobic polar compounds which have low permeability through hydrophobic polymer such as silicon rubber or EVA membrane.

After studying the controlled release of scopolamine drug through EVA membrane in transdermal patch formulations, release rates were compared with uncontrolled reservoirs. Then, it was found that an EVA membrane patch released scopolamine at a constant rate for more than 72 hours.

Krishna and pandit (1994) prepared three transdermal formulations containing propranolol hydrochloride in a hydrophilic polymer matrix in three different forms like one without rate controlling membrane and other two with EVA rate controlling membrane of different thickness. It was found that increased thickness of EVA ultimately led to greater retention of the drug in device and zero order profile was observed with EVA.

Rate controlling membrane may be prepared by either solvent evaporation method or compression method.

1. In solvent evaporation method, polymer is dissolved in solvent with or without plasticizer followed by pouring the solution on the horizontal surface and left for evaporation of solvent in order to obtain a thin film.
2. In case of compression method, polymer is compressed with required force for specific time at high temperature. Drugs such as testosterone which require relatively high doses of greater permeation enhancement, then liquid reservoir systems are used. But the application of enhancers and adhesive technologies has made it possible to formulate many drugs that were initially administered in liquid reservoirs to be used as matrix type systems. Examples are oestradiol, nicotine, nitroglycerine. The main advantage of reservoir type patches is that this patch design can provide a true zero order release pattern of drug so as to achieve a constant serum drug level. E.g., Duragesic, estradem and androderm.

Membrane matrix hybrid type patch

This patch is the modification of reservoir type transdermal patch. In this, the liquid formulation of the drug reservoir is replaced with a solid polymer matrix (polyisobutylene) which is sandwiched between rate controlling membrane and backing laminate or layer. E.g., Catapress and transdermscop.

Micro reservoir type transdermal patch

In this, the drug reservoir is formed by suspending the drug solids in an aqueous solution of water miscible drug solubiliser. Example is PEG. The drug suspension is homogenously dispersed in lipophilic polymer such as testosterone, thus forming thousands of unreachable microscopic drug reservoirs (micro reservoirs). The dispersion formed is quickly stabilized by immediate formation of cross linkage between the polymer chains *in situ* which produces a medicated polymer disc of a specific area and fixed thickness. Occlusive base plate which is mounted between the medicated disc and adhesive form backing prevents the loss of drug through the backing membrane. E.g., Nitrodisc.

Drug in adhesive type transdermal patch

In this, the drug and other selected excipients are directly incorporated into the organic solvent based pressure sensitive adhesive solution. Then it is mixed, cast as a thin film and dried finally to evaporate the solvents leaving a dried adhesive matrix film containing the drug and excipients. This drug in adhesive matrix is sandwiched between release liner and backing laminate. Drug in adhesive patch may be either as a single layer or multi layer. The multilayer systems arc different from the single layer in that and addition of another layer of drug in adhesive usually separated by a membrane. Some examples of suitable pressure sensitive adhesives are polysilaxanes, polyacrylates and polyisobutylene. These are hydrophilic in nature and are prepared as solutions of polymer dissolved in organic solvents. So, this type of system is preferred for hydrophobic drugs because of the reason that it is to be incorporated into organic solvent based hydrophobic adhesive.

Rachel prepared drug in adhesive patches of green tea extract and it was observed for the delivery of major catechins and caffeine extracted from green tea successfully through this transdermal patch from drug in adhesive patches.

Evaluation of Transdermal Patches

Development of controlled release transdermal dosage form is a complex process. Transdermal patches have been developed in order to improve clinical efficacy of the drug and to enhance patient compliance by delivering smaller amount of drug at a predetermined rate. This makes evaluation studies even more important in order to find out their desired performance and reproducibility under the specified environmental conditions.

The evaluation studies performed on transdermal patches were classified as follows

- Physicochemical evaluation
- *In vitro* evaluation
- *In vivo* evaluation

Physico Chemical Evaluation

Thickness

The thickness of transdermal film is measured by traveling microscope, dial gauge, screw gauge at different points of the film.

Uniformity of weight

Weight variation is studied by individually weighing 10 randomly selected patches and the average weight is calculated. It is found that the individual weight should not deviate from average weight.

Drug content determination

An accurately weighed portion of film (100 mg) is dissolved in 100 ml of a suitable solvent in which drug is soluble and then the solution is shaken continuously for 24 hours in shaker incubator. Then the whole solution is solicited. After fornication and subsequent filtration drug in the solution is estimated by using a spectrophotometer by appropriate dilutions.

Drug content uniformity test

Ten patches are selected randomly and content is determined for individual patches. From the results, if 9 out of 10 patches have content between 85-115% of the specified value and one has content not less than

75-125% of the specified value, then the transdermal patches pass the drug content uniformity test. But if three patches have content in the range of 75-125% then additional 20 patches are tested for drug content. If these 20 patches have range from 85-115% then the transdermal patches are found to pass the test.

Moisture content

The prepared films are weighed individually and kept in desiccators containing calcium chloride after a specified time of 24 hours at room temperature. These films are weighed again after a specified interval until they show a content weight. The percent moisture content is calculated by using following formula.

$$\%\text{ moisture content} = \frac{\text{initial weight} - \text{final weight}}{\text{final weight}} \times 100$$

Moisture uptake

First weigh the films and then the films are kept in desiccators at room temperature for 24 hours. These are then taken out and exposed to 84% relative humidity using saturated solution of potassium chloride in desiccators until a constant weight is achieved. The percentage of moisture up take can be calculated by using following formula.

$$\%\text{ moisture uptake} = \frac{\text{final weight} - \text{initial weight}}{\text{initial weight}} \times 100$$

Flatness

This study is being performed because the transdermal patch should possess a smooth surface and should not constrict with time. For, flatness determination one patch is cut from the centre and two from each side of patches. The length of each patch is measured and variation in length is measured by determining percent constriction. 0% constriction is equivalent to 1005 flatness.

$$\%\text{ constriction} = \frac{I_2 - I_2}{I_2} \times 100$$

I_1 = initial length of each strip

I_2 = final length of each strip

Folding endurance

Evaluation of folding endurance involves determining the folding capacity of the films after subjecting them to frequent extreme conditions of folding. Folding endurance is determined by repeatedly folding the film at the same place until it breaks. The number of times the films could be folded at the same place without breaking is folding endurance value of that particular patch.

Tensile strength

To determine tensile strength, the polymeric films are sandwiched separately by corked linear iron plates. In this test, one end of the films is kept fixed with the help of an iron screen and other end is connected to a freely movable thread over a pulley. The weight is added gradually to the pan that is attached with the hanging end of the thread. A pointer on the thread is used to assess the elongation of the film. The weight which is just sufficient to break the film is noted. The tensile strength can be calculated by using the following equation.

Tensile strength = F/alb (1+L/I)

F = Force required to break

A = Width of film

B = Thickness of film

L = Length of film

I = Elongation of film at break point.

Microscopic studies

Distribution of drug and polymer in the film can be studied by using scanning electron microscope. For this study, the section of each sample is cut and then it is mounted on to stubs using double sided adhesive tape. The sections are then coated with gold palladium alloy by using fine coat ion sputter so as to render them electrically conductive. Then the sections are again examined under scanning electron microscope.

In vitro Release Studies

Drug release mechanism and kinetics are the two characteristics of dosage forms. These play an important role in describing the dissolution profile from a control release dosage forms.

There are various methods available for determination of drug release rate of TDDS.

- ***The paddle over disc:*** This method is identical to USP paddle dissolution apparatus except that the transdermal system is attached to a disc or cell resting at bottom of the vessel that contains medium at 32°C.
- ***The cylinder modified USP basket:*** This method is similar to the USP basket dissolution apparatus expect that the system is attached to the surface to hollow cylinder which is immersed in medium at 32°C.
- ***The reciprocating disc:*** In this method, patches attached to holders and are oscillated in small volume of medium allowing the apparatus to be useful for systems delivering low concentration of the drug. In addition to this, paddle over extraction cell method may be used.

Diffusion Cells

For example, Franz diffusion cell and its modification Keshary-Chein Cell. In this method, transdermal system is placed between receptor and donor compartment of the diffusion cell. The transdermal system should face the receptor compartment in which receptor fluid i.e., buffer is placed. The agitation speed and temperature should be kept constant. The whole assembly is kept on magnetic stirrer and solution in the reservoir compartment is constantly and continuously stirred throughout experimental time by using magnetic beads. At predetermined time intervals, the receptor fluid is removed for analysis and is replaced again with an equal volume of fresh receptor fluid. The concentration of the drug determined by using spectrophotometer.

The pH of the dissolution media ideally should be between 5-6. The test temperature is 32°C at 100 rpm. The dissolution data that is obtained from the test is fitted to mathematical models in order to ascertain the release mechanism.

In vitro Permeation Studies

The amount of drug that is available for absorption to the systemic pool is greatly dependent on drug released from the polymeric transdermal film. The drug which reaches at skin surface is then passed to usual permeation studies which are performed by placing the fabricated transdermal patch attached to rat skin or synthetic membrane in between receptor and donor compartment in a vertical diffusion cell such as Franz diffusion cell or keshary-chine diffusion cell. The transdermal system is applied to the hydrophilic side of the membrane and then it is mounted in the diffusion cell with lipophilic side in contact with receptor fluid. The receiver comportment should be maintained at specific temperature usually 32°C and is continuously stirred at a constant rate. The samples are withdrawn at equal time intervals and equal amount of buffer is replaced each time. The samples are diluted appropriately and absorbance is determined by using spectrophotometer. Then the amount of drug permeated per square centimeter at each time interval is calculated. Design of system, patch size, surface area of skin, thickness of skin and temperature etc. are some of the variables that may affect the release of drug. So, permeation study involves many steps like the preparation of skin, mounting of skin on permeation cell, setting of experimental conditions like temperature, stirring, skin conditions, withdrawing samples at different time intervals, sample analysis, and calculation of flux.

In vivo Studies

In vivo evaluation is the true depiction to assess the drug performance. The variables that cannot be taken into account during *in vitro* studies can be assessed during *in vivo* studies. *In vivo* evaluation of TDDS can be carried out by using

- Animal models
- Human volunteers

Animal models

Animal studies are preferred mainly because considerable time and resources are required to carry out human studies. The most common animal species which are used for evaluating transdermal drug delivery system are mouse, hairless rat, hairless dog, hair less rhesus monkey rabbit, guinea pig etc. Various experiments conducted on these animals lead us to a conclusion that hairless animals are preferred over hairy animals in both *in vitro* and *in vivo* experiments. The most reliable model used in these tests is Rhesus monkey for *in vivo* evaluation of transdermal drug delivery in man.

Human models

The final stage of the development of a transdermal device involves the collection of pharmacokinetic and pharmacodynamic data from human volunteers by the application of the patch. Clinical trials have been conducted to measure various parameters like the efficacy, risk involved, side effects, patient compliance etc. Phase I clinical trials are conducted to determine mainly safety in volunteers and phase II clinical trials determine short term safety and mainly effectiveness in patient. Phase III trials indicate the safety and effectiveness in large number of patient population and phase IV trials at post marketing surveillance are done for marketed patches to detect adverse drug reactions. Though human studies require considerable resources and are problematic but they are the best to assess the performance of the drug.

Skin Irritation Studies

White albino rats, mice or white rabbits are generally used to study any hypersensitivity reaction on the skin. Mutalik and Udupa (2005) carried out skin irritation test using mice. The mice were divided into 5 groups, each group with 6 animals. The hair on the backside area of mice was removed on the previous day of the experiment. The animals of group 1 are considered as normal, without any treatment. One of the group of animals (group 2, control) were applied with marketed adhesive tape. Transdermal systems (both blank and drug loaded) were applied onto nude skin of animals of the groups 3 and 4. The standard irritant which is 0.8% v/v aqueous solution of formalin was applied on the animals of group 5. New patch is being applied on the animal each day up to 7 days and finally the application sites were measured according to a visual scoring scale, always by the same investigator. The erythema found on the skin as follows: 0 for none, 1 for slight, 2 for well defined, 3 for moderate and 4 for severe. After visual evaluation of skin irritation, the animals were sacrificed and then collect skin samples which were further processed for histological examination. The results of this study are found to be like the prepared systems (blank and drug loaded) and USP adhesive tape produced negligible erythema and edema. The histological examination of the skin also indicated that adhesive tape and prepared patches produced only mild inflammation and edema.

Stbility Studies

The stability studies are conducted in order to investigate the influence of temperature and relative humidity on the drug content in various formulations. These transdermal formulations are subjected to stability studies as per IC guidelines.

Applications of Transdermal Patches

- **Transdermal patches for ischemic heart disease**

  The transdermal patch of isosorbid dinitrate in a viscoelastic polymer matrix is used to treatment of angina pectoris and other ischemic heart diseases. When applied to the chest the patch releases the drug through reservoir system thereby contributing to stable improvement of cardiac functioning.

- **Transdermal patches for asthma**

  It is world first adhesive patches formulation that contains tolbuterol usually used for treating asthama. Generally it is known that there is a rhythm in respiratory function which declines from midnight to early morning. This product is perfect for such condition "take medicine when you need with the amount you need".

- **Patch for local anesthesia**

  This transdermal patch contains lidocaine, which is usually used to anesthetize skin, and relieves the pain experienced on insertion of large needles used in hemodialysis. The pain from such needles increases fear and distress and clearly reduced the medical benefits, nitto denkols drug delivery technology introduces a painless puncture that delivers a local anesthetic into the skin at a high concentration.

- **Commercial transdermal systems**

  At present the drugs give in combination patches are approved for administration via transdermal route. These contain scopolamine, nitroglycerin, nicotine, diclofenac, estradiol and testosterone etc.

Nitroglycerins

It is a vasodilator used in treatment of angina pectoris.Transdermitro is a membrane modulated system were as nitrodur and nitrodisc are based on the polymer matrix are devised the later being a macro sealed drug delivery system. In these system the nitroglycerin is normally as a lactose triturates.

Nitro_Dur System

In this nitroglycerine is distributed homogenously throughout a polymer matrix containing polyvinyl pyrrolidone and polyvinyl alcohol. These are releases nitroglycerin linearly with the square root of time and for a 10 cm^2 patch containing 51 mg, 80-90% of drug is released over 24 hours.

- **Estradiol**

  The post menopausal syndrome is characterized by a variety of symptoms including flushes, atrophic vaginics, sleep disturbance and alterations of secondary sexual charecteristics. Approximately 40% of women develop post menopausal syndrome. This estrogen replacement therapy provides mainly elevated hepatic proteins and estrogen level. This effects reduced by avoiding first pass metabolism. The suggested route is transdermal route.

  This patch is designed for twice week application to 7 days application.

- **The patch for food**

  For years smokers looking to quit have used "the patch" to deliver small doses of their nicotine fix. Now the army is working on a similar systems to give soldiers the nutrients they need during war time.

  Here how the transdermal nutrient delivery system work is tiny physiological monitors would keep tabs on a soldier's metabolism and then send information on his nutritional needs to a processor. This processor would then trigger a micro electrical mechanical systems that would transmit the micronutrients are pumped directly into blood capillaries into soldier's body.

  But the Natick soldier center says it won't be ready for another decade.

- **Transdermal beauty patch**

  Say good bye to creams. The new cosmetic trend is transdermal patches. Because they use our skin effectively, continuously, deliver their beauty agents into our skin and organism. They also serve to tackle bags under the eyes, eliminate facial blotches and shape and hydrate skin.

  These are very comfortable and easy to carry, quickly to apply and show visible results much faster.

- **Patches enter in the cosmetic world**

  After their success in the pharmaceutical environment the patch entered the world of cosmetic. The first patch, not yet transdermic ones used a sanitary gauze or paper for support.

INDEX

F

G

H

I

L

M

N

O

P

S

T